The Effect of Modern
Agriculture on
Rural Development

Pergamon Titles of Related Interest

Avery RURAL CHANGE AND PUBLIC POLICY
Barney THE GLOBAL 2000 REPORT TO THE PRESIDENT OF THE US
Ewell/Poleman UXPANAPA: Agricultural Development in the
  Mexican Tropics
Golany ARID ZONE SETTLEMENT PLANNING
Goodman/Hawkins/Miyabara FOOD AND AGRICULTURAL WASTE
  DEVELOPMENT
Lonsdale/Holmes SETTLEMENT SYSTEMS IN SPARSELY
  POPULATED REGIONS
Miljan/Laszlo/Kurtzman FOOD AND AGRICULTURE IN
  GLOBAL PERSPECTIVE
Volgyes/Lonsdale/Avery THE PROCESS OF RURAL
  TRANSFORMATION

Related Journals*

AMBIO
CURRENT ADVANCES IN ECOLOGICAL SCIENCES
ENVIRONMENT INTERNATIONAL
JOURNAL OF THE AUSTRALIAN INSTITUTE OF AGRICULTURAL
  SCIENCE
SOIL BIOLOGY & CHEMISTRY

*Free specimen copies available upon request.

PERGAMON
POLICY
STUDIES
ON INTERNATIONAL DEVELOPMENT

The Effect of Modern Agriculture on Rural Development

Edited by
Gyorgy Enyedi
Ivan Volgyes

Comparative Rural Transformation Series

Pergamon Press

NEW YORK • OXFORD • TORONTO • SYDNEY • PARIS • FRANKFURT

Pergamon Press Offices:

U.S.A.	Pergamon Press Inc., Maxwell House, Fairview Park, Elmsford, New York 10523, U.S.A.
U.K.	Pergamon Press Ltd., Headington Hill Hall, Oxford OX3 0BW, England
CANADA	Pergamon Press Canada Ltd., Suite 104, 150 Consumers Road, Willowdale, Ontario M2J 1P9, Canada
AUSTRALIA	Pergamon Press (Aust.) Pty. Ltd., P.O. Box 544, Potts Point, NSW 2011, Australia
FRANCE	Pergamon Press SARL, 24 rue des Ecoles, 75240 Paris, Cedex 05, France
FEDERAL REPUBLIC OF GERMANY	Pergamon Press GmbH, Hammerweg 6 6242 Kronberg/Taunus, Federal Republic of Germany

Library of Congress Cataloging in Publication Data
Main entry under title:

The Effect of modern agriculture on rural develop-
    ment.

    (Pergamon policy studies on international
development)
        Includes index.
        1. Rural development--Case studies.
2. Agriculture innovations--Case studies.
3. Land settlement--Case studies. I. Enyedi,
Gyorgy. II. Volgyes, Ivan, 1936-   III. Series.
HD1415.E33   1982   338.9'009173'4   80-25232
ISBN 0-08-027179-0

Printed in the United States of America

Contents

PART II: MODERNIZING AGRICULTURE AND THE RURAL SETTLEMENT PATTERN

PART III: AGROTECHNIQUES AND RURAL CHANGE

Introduction

Gyorgy Enyedi
Ivan Volgyes

The problems of rural development are increasingly of interest to regional planners and regional policymakers. Rural development projects in the developed countries, and often in the developing countries as well, put more emphasis on rural industry, tourism, or services than on the modernization of agriculture. But the latter still remains the basic rural function, at least in terms of resource utilization. Agriculture is not generally treated as a dynamic element of regional development. The reason is simple: the rapid development of agriculture, contrary to the rapid growth of industry or the tertiary sector, has generally been accompanied by a substantial decline of employment.

Consequently, at least according to the experiences of the developed nations, modern agriculture cannot fulfill the basic function of a developed region, because depopulation undermines the functioning of services.

These effects are of a certain type of technical-economic modernization. On a global scale, however, we must suppose the existence of different types, aspects, and, consequently, different effects of agricultural modernization.

For what we call "modern agriculture" today, North America has served as the innovation center. This type of modern agriculture has been characterized by high capital intensity, a high technical level, massive utilization of machinery, clear profit-orientation, very low manpower use, owner-operated private farms, and a great abundance of

land. We can add that agriculture has played a
secondary role as compared to the dominant industry,
and that the density of agricultural population was
already low at the beginning of modernization. This
North American type of technical modernization was
introduced, with modifications, to Western Europe.
The western part of the European continent has a
long tradition in certain agricultural innovations,
such as animal breeding and crop rotations, but the
highly mechanized agricultural technology was transplanted there from North America. Modifications were
justified by the smaller scale of farms and by
certain peasant traditions surviving in Western
Europe. A special innovation center for modern
agriculture was formed in Japan, suitable for the
microfundia and high density of agricultural population.

Despite the important differences existing between North America, Western Europe, and Japan,
there are fundamental common features: the profit-
orientation of farming, the primacy of the economic
goal to social aspects, and the ability of secondary
and tertiary sectors to absorb the surplus manpower
leaving agriculture. The North American type of
agricultural modernization has one rather negative
effect upon rural development - depopulation. The
utilization of modern technology has certain advantages, such as the development of the rural infrastructure and a higher education of the agricultural
population, but in the long run, the rural depopulation might have disastrous effects.

Consequently, this North American type of modernization might be adapted well to other continents
only in cases of similar social structure and
production goals, such as in Israel or South Africa.
But its introduction into the very different rural
societies of developing countries has often had
contradictory or even destructive results, as illustrated by a great number of case studies. Every
basic type of rural society needs its own kind of
agricultural modernization tailored to express different aims and methods.

The East European socialist countries have
their independent type of agricultural modernization. Its most important characteristics are as
follows:

- Dominance of collective (cooperative or
  state-owned) large-scale farming.

- Relatively high technical standards (technology partly adapted from the North American innovation center).

- The food supply of the country's population might have priority among production goals.

- The decline of agricultural employment should not harm full employment.

Large-scale farms in Eastern Europe are economic as well as social organizations. They offer a number of services, offered elsewhere through village or regional organizations, to the rural population. Large-scale farms, playing an important role in vertical integration, can attract new commercial and food-processing activities to the countryside. In this way, large farms can increase employment in their nonagricultural agribusiness activities, and, at least in the important agricultural regions, they can more or less stabilize the rural population. Nonetheless, it must be pointed out that the East European model includes the experiences of economies (such as Yugoslavia and Poland) where private farms remain dominant as well as the experiences of the other socialist states where the productive activities are dominated by the state or collective sectors.

Agricultural modernization in developing countries has to be based on social modernization, that is, agrarian reforms. First of all, a new social structure that makes the producers interested in economic growth and which is able to absorb technical innovations must be formed. The technical innovations have to be oriented toward higher yields, and by no means toward a sharp diminishing of manpower use. It is clear that the uncontrolled growth of large cities in developing countries, caused partly by the immigrations of rural surplus manpower, is menacing. Thus, the developing countries have to modernize their agriculture in order to ease the food shortage, but also have to try to enlarge rural employment. Contrary to the trend in developed nations, the agricultural modernization of less developed countries is taking place before a general industrialization.

This type of agricultural modernization also has a number of nonproductive goals, such as developing rural infrastructure, basic health services, and education. One can base the modernization on the existing social organization of rural communities. The experiences of the people's communes of China are worth studying; these organizations, despite the frequent political upheavals in China, were successful in feeding almost a billion people and assuring basic services for the rural population. One can expect the formation of other special modernization types within the developing world.

It is well known that the former "underdog" role of food and agricultural raw material production has substantially changed in the world economy. Food has become an important strategic and political tool as well. The role of agriculture has to be reevaluated also in national development policies. Governments have to recognize that agricultural modernization is mostly a social phenomenon, and open the way for social transformation. Hence, agricultural modernization can lead to the comprehensive transformation of the rural society and the rural settlement network; the integration process of modern agriculture can enlarge the rural industry and rural services. Consequently, we can expect that modern agriculture will become, at least, one of the major dynamic elements of development in the productive regions of the earth.

This volume in the series on comparative rural development deals with the role of agriculture in the process of rural transformation. It attempts to examine the interchange between the social and technical elements of change by focusing on the alteration of societal and settlement changes, on the one hand, and on the alterations of technology, industrialization, and integration of production, on the other. The authors of the individual chapters address these topics by trying to illustrate how the various processes of change have affected the experiences of various states at all levels.

This book is divided into five sections. The first section deals with social change, the modernization of agriculture, and the general process of rural transformation. The studies selected focus on the examples of three advanced economies (Canada, Belgium, and Australia), on the centrally directed economy of East Germany, and on the rapidly developing economy of Puerto Rico. Although the processes and the levels of development are as different as the speed and the direction of change, the authors of chapters in this section all focus their attention on the social change attendant with the process of rural transformation.

In the second section, the authors concentrate on the changes in the settlement process as the modernization of industry begins to turn from extensive to intensive uses of manpower and as society strips the excess manpower for industrial use. The focus is on the variety of change and the divergent experiences in settlement structures, ranging from rural depopulation to fairly stable rural population patterns. The studies stress the close connection between modernization and urbanization, but make it clear that the process of urbanization is not and need not be synonymous with the process of the general modernization and industrialization of societies. The section includes studies dealing with the United States, France, Finland, the USSR, Poland, and Hungary in order to illustrate the universality of problems in spite of obvious systemic and development differences.

The third section examines the effects of changing agricultural technology on a modernizing countryside. It attempts to illustrate that new methods and processes in agricultural technology bring not merely heightened levels of expectation, but a change of attitudes as well. A variety of technological changes are examined in this section: changes in dairy, fruit, and viniculture, as well as the introduction of specialized crops. The authors in this section deal with such diverse examples as Canada, Japan, Israel, the United States, and Hungary to illustrate that the introduction of modern agro-techniques, regardless of the level of the general development of the economy, always results in a significant improvement in production levels.

Just as technological change is important, an attendant, though less clearly delineated and more complex alteration, can also be witnessed in the changing countryside, namely, the general industrialization of agriculture. This process is, of course, a relatively new phenomenon in the changing environment of rural life, and it combines the activities of the primary sector with those of the secondary sector. Through such activities as the location of industrial plants in the rural sphere, regardless of the differences in political systems, modernizing states are able to create year-round employment for the rural population, utilize excess or partially excess work forces, including the under-utilized labor of women, and induce economic planners to combine the two, traditionally separate

productive sectors. As the authors of this section
illustrate through such examples as Hungary and
Israel, industrialization of agriculture is especially viable in small states where urbanization
and industrialization are not coterminous and where
manpower scarcities generally exist.

The concluding section of this volume is
devoted to the process of agro-industrial integration and the attendant changes of the rural sphere.
As modern systems of production are introduced and
as the industrial and rural spheres are intertwined
for both economic and regional reasons, the necessity for cooperation grows. Focusing on the United
States, West Germany, and Hungary, the authors of
this section illustrate the various forms of
agro-industrial cooperation and integration and the
relationship between such activities and the process of rural change.

The chapters for this volume were specifically
commissioned and were delivered in paper form first
at the Fourth International Meeting organized by
the Commission on Rural Development of the International Geographic Union. They were orally delivered in Szeged, Hungary in 1979 and edited at the
University of Nebraska, under the auspices of the
Graduate Program on Comparative Rural Transformation of the University. Partial assistance for the
typing and editing of the manuscripts was given by
Title V of the Rural Development Act. That assistance is gratefully acknowledged by the editors of
this volume.

The typesetting for this volume was done by
Donna Pett. Her careful and conscientious work made
it possible for us to complete this volume. The
editors wish to express their gratitude for her
contribution and express their appreciation for her
efforts.

I
Social Changes, the Modernization of Agriculture, and the Process of Rural Transformation

1 The Impact of Modern Agriculture on Rural Transformation in Canada

Lewis A. Fischer

Extending as it does from the Atlantic to the Pacific, from the frozen territories of the Arctic to the temperate zones of the Niagara Peninsula, Canada's vastness embodies a great variety of natural conditions, a multiformity of societal groupings, and a diversity of economic opportunities. Like the majority of industrialized countries, Canada has a mixed economy. It neither follows the classical type of capitalism nor is a centrally planned system. It is rather a conglomerate of market-oriented subsystems, with different levels of competition. Pricing occurs in the framework of business enterprises, yet a substantial part of production is under government control. The government does not control the quantity of goods produced, but it has a host of instruments to exercise supply management. The decision-making power on agricultural policies and programs is divided between one federal and ten provincial governments. Inevitably, provincial governments may frequently represent fundamentally divergent regional interests.

Against this background, this chapter identifies some major characteristics of the modernization process in Canadian agriculture and assesses their impact on rural development.

SELECTED INDICATORS OF MODERNIZATION IN AGRICULTURE

(a) Natural Resources

Canada has some 904 million hectares (2.200 million

1

acres) of land, of which only 7 percent (67 million
hectares) is in farming. As indicated at the
outset, a wide range of climates exists over the
agricultural regions of Canada. The frost-free
period extends from minimum of 80 days to the
maximum of 220 days, the latter being restricted to
some coastal areas. Appendix 1.1 shows that according to recent estimates, some 114 million hectares
of land could be used for agricultural purposes;
that is almost twice the present farming area. Out
of 43.7 million hectares of "improved land," 32
million hectares are under crops and improved
pasture. The average amount of improved land per
farms has increased significantly from 30 percent
in 1930 to 65 percent in 1976. (1) This demonstrates (a) that there is potential for expansion
of production, (b) that intensity and efficiency
could be improved if financial resources would be
made available, and (c) that Canadian farmers have
both the skill and the desire to intensify farm
operations.

An important feature of development is the
increasing competition for farmland from industrial
and urban expansion. Since the complex network of
medical, educational, and cultural services in
rural areas are seriously endangered by this trend,
it must be recognized as a force affecting both
farm and nonfarm rural conditions. Accurate data
on the rate of loss of farm land to urban use are
not available, but Agriculture Canada estimates
that the area absorbed for every increase of 1,000
in urban population varies from about 80 to 400
hectares. The higher figure includes urban
fringe--land alienated from agriculture by land
speculation and resulting high prices. If the value
of 80 hectares per 1,000 increase of urban population is used, the projected permanent conversion of
land to urban development in the provinces Quebec
and Ontario between now and the year 2000 totals
300,000 hectares and 500,000 hectares, respectively. (2)

(b) Human Resources

Changing agricultural structure is a worldwide
phenomenon characterized by a decreasing farm population, increasing average farm size, and an adjustment in resource use. Technological progress

is a major factor reflected in the substitution of
capital for human or animal labor resulting in a
shifting investment toward purchased inputs. The
way farmers perceive technical and biological innovations is central to their decision-making process. In accordance with its prime objective, this
study's main focus is the change in the human,
social, and economic characteristics of the modern
farm industry. During the period 1961-1976, the
number of farm operators declined from 480,000 to
300,000; that is, by 37.5 percent. In contrast to
European trends, the number of hired laborers
decreased at a much lower rate, and even showed
some increase in 1974 and 1975. During the period
1951-1971, the total farm population declined from
2.9 million to 1.5 million, falling from 20.8 to
6.9 percent of the total population. These decreases have had a significant impact on the
character of many rural communities. (3) The reductions occurred in different regions with differing
extents and rates. (4)

Most farms in Canada are operated by their
owner. Recently, however, there has been a trend
toward corporate companionship between family members. As the proportion of "legally constituted
incorporated family businesses" increases, these
provide improved credit and tax preferences for
their members. In a similar vein, the system of
"partnership" between father and son is slightly
decreasing. Table 1.1 reveals the type of organization in 1976.

Table 1.1. Type of Farm Organization in Canada
1976

Type of Organization	Number of Farms	Percentage of Total
Individual/family	274,490	91.5
farm Partnership	11,313	3.8
Institution/community	771	0.3
Legally constituted		
company: Family	11,348	3.8
Other	1,864	0.5
Other types	332	0.1
TOTAL	300,118	100.0

Source: Statistics Canada, Cat. 96-800, 1976,
p. 28-Table 1

Appendix 1.2 demonstrates the transformation of the agricultural labor force in Canada under the impact of modernization. The most striking feature is the negative relationship between the number of unpaid family workers and hired workers. Unpaid family workers have traditionally played a major role in farm production, outnumbering paid workers until the mid 1970s when a shift towards a greater reliance on hired labor began. (5) The social effect of this trend on rural development is significant in that it changed the structure of the farm family as young family members opted for industrial jobs. Regular wages, regular working hours, and better access to recreational centers have attracted many of them to urban areas. Some return, however, frustrated by the lack of opportunities for inexperienced and inefficient job aspirants. But indeed, most farmers now prefer seasonal workers anyway. (6) But in so doing, they ensure a pool of unemployed labor for much of the year. All these changes created increasingly complex labor relations in rural areas in the early 1970s. More recently, the situation has eased as minimum-wage schemes and special government sponsored farm labor programs have been implemented.

(c) Capital Resources, Allocations, and Returns

Technological change is an important element of modernization. Theoretically, it is the transfer of scientific discoveries to agriculture, while in practice it is often the introduction of labor-saving methods to farm operations. This process of change depends greatly on the nature of the innovations. In Canadian agriculture the dominant characteristic has been the increased substitution of capital for labor.

Advances in transportation and the replacement of human and animal labor by mechanical power have had profound effects on the social structure of the rural farm and nonfarm population. The scope of this study does not permit a cataloging of all the elements of these trends. The following section is limited to the major components of technological change.

Table 1.2 reveals that, parallel to the decline of farm numbers, the number of tractors, grain combines, and trucks has increased. It further shows the growing volume of capital invested

in machinery. (7) Also, the consumption of fertilizers by Canadian farmers increased in the 1950-
1975 period from 193,000 tons of the three primary
nutrients (nitrogen, phosphorus, and potassium) to
1,232,000 tons. Expenditures on fertilizers and
limestone grew from $40.3 million to $497 million.

Table 1.2. Number and Distribution of Selected
      Farm Machinery in Canada 1952-1976

	1951		1976	
	Number of Machinery on Farms	Farms Reporting	Number of Machinery on Farms	Farms Reporting
Autos	252,610	240,120	291,223	236,385
Tractors	331,587	278,285	605,284	275,241
Grain Combines	83,328	79,580	161,852	142,344
Trucks	154,445	142,298	422,598	236,815

Source: Statistics Canada, Cat.96-800, p.61

     The above examples of technological change
illustrate the trend of modernization in Canada.
The major goal of this trend was to improve
efficiency and to promote greater productivity.
Increasing capital input resulted in higher average
yields per land unit. Indeed, fewer farmers now
grow more food on less land. The number of persons
supplied with food and fiber by the production of
one farm worker was 11 in the 1930s, changed to 25
during the second half of the 1950s and was 41 by
1977. Total cash receipts during the period 1963-
1972 averaged $4,192 million, reached $6,741 million in 1973, and was $11,900 million by 1978. (8)
     In general, modern Canadian agriculture is a
conglomerate of commercial enterprises and is
clearly market oriented. There is a strong trend
toward further modernization. However, the basic
aspiration of the farmers is the stabilization of
rural living conditions and the preservation of the
family farm system.

                    THE FAMILY FARM

The transformation of economic and social conditions that has taken place in Canada's national
economy in the past twenty-five years presents the

greatest challenge to traditionally family farm-oriented agriculture. Modernization has made farming an extremely complicated enterprise. It is postulated that one farmer combines the basic knowledge of a mechanic, an agronomist, a veterinary surgeon, and a manager, along with adequate marketing skill. The fundamental characteristics of the family farm in Canada are the ownership of production factors and the freedom of decision making. On the road to modernization, the farmer has encountered conflicting values and ideologies embodied in the concept of the viability of the family farm. The crucial dilemma is expressed in two contradictory slogans: "grow or go" and "small is beautiful".

The traditional classification of farm size in hectares indicates that from 1951 to 1976 the average farm size increased by 53.5 percent, from 145.8 to 223.4 hectares. However, in the modern farm industry, the size of a farm cannot be regarded as the sole indicator of viability or efficiency. In many sectors, specialization and integration have created farm enterprises with large investments and large business capacities located on relatively small pieces of land. Poultry production, feed-lot operations, and several other business practices are examples of so-called landless farming. Hence, a distinction should be made between the size of a farm and the size of an operation. Nonetheless, Table 1.3 reveals that the increase in the value of products sold is in direct proportion to the increase of average farm size. The reason for this is the relatively small number of "landless" operations.

Using the material in Table 1.3, it is possible to divide Canadian farmers into three economic groups: one group, representing 35 percent of farmers, is viable (sales $25,000 and over); another group (42 percent) is neither well-off nor poverty stricken; and a third group is below the poverty level. The size of the poor or "marginal" farm sector helps explain the increase in the importance of off-farm work in the last twenty-five years. The proportion of farm operators reporting off-farm work was 28 percent in 1951 and 35.3 percent in 1971.

Table 1.3.   Census Farms Classified by Value of
           Agricultural Products Sold - 1976

Value of Agricul- tural Products Sold	Number of Farms	Percent of Total	Average Farm Size (hectares)
$ 1,200 - $ 4,999	69,097	23.0	88.9
$ 5,000 - $ 9,999	45,791	15.2	132.1
$10,000 - $24,999	81,492	27.2	197.6
$25,000 - $49,999	59,309	19.8	261.6
$50,000 - $74,999	22,120	7.4	265.6
$75,000 - $99,999	9,189	3.1	416.1
$100,000 & over	12,349	4.1	620.1
Institutional farms	771	0.2	
TOTAL NUMBER	300,118	100.0	223.8

Source: P. M. Moncrieff et al, Canada's Agricultural System, 3rd ed., (Montreal: Macdonald Col-
lege, 1978), p.3.

Off-farm incomes have contributed significantly to
the economic welfare of farmers. Data show that
while the farmer's total income from all sources
increased from $3,606 in 1958 to $10,018 in 1974,
the share of farm income declined from 65 percent
to 44 percent. Regional discrepancies are responsible for regional diversity in the relative significance of farm and off-farm income. (9) The
widespread practice of part-time farming is relatively new, but is now a rapidly expanding part of
the rural adjustment process, generating the major
part of off-farm income. This phenomenon is widely
encouraged in some West European countries; its
social and economic significance has been only
recently recognized in North America.
        Improvements in marketing technology have also
contributed to modernization by reducing marketing
costs. Compulsory bulk transportation of milk,
controlled atmosphere storage, dehydration, truck
transportation, and many other technological innovations are all components of the modern market
process. The majority of family farms has gained
from these developments. A minority, however, has
faced severe challenges of adjustment which have
frequently exceeded their financial capacity.

Innovations in production and marketing have encouraged both vertical and horizontal integration. The latter is reflected in enlarged farm sizes as farmers sell or rent their land to neighbors or as previously independent family members pool their resources. The growth of vertical integration, or contract farming, signals specialization in both production and marketing. It may become the vehicle of survival for the family farm, provided that it includes the dissemination of knowledge, freedom in decision making, and freedom to terminate the contract. Despite the potential for contract production, the trend in recent years has been to discourage this phenomenon with the establishment of marketing boards. There are now more than one hundred producers' marketing boards in Canada with authority to fix prices and control supplies. Their essential task is to protect the farmer's economic interest.

The cooperative movement is limited to two sectors in two regions. The grain farmers of western Canada developed a large and efficient cooperative organization which became a vital element of the grain-handling system. In the dairy sector, a sweeping tendency toward concentration has eliminated the small processing units in Ontario and Quebec; the present system is an oligopoly with farmers' participation. In other production domains, the cooperative movement is scarcely growing in Canada. Consumer's cooperatives organizing wholesale buyer units and running their own retail stores, service stations, medical clinics, and funeral centers have difficult times, even in rural areas, competing with the expanding modern chain-corporation.

In the years ahead, we can expect to see in Canada a continuation, even an acceleration, of the trend toward modernization of agriculture. The formative effect of this trend on the family farm is fundamental and takes many forms. There is a general consensus that the family farm is still the basis of a desirable pattern for agricultural and rural living in the country. The real answer to the survival of the Canadian family farm is neither to get big a tout prix nor to refuse the challenge of efficiency if it calls for running an enlarged scale enterprise. Farmers must bring production into line with demands in accordance with their own physical, educational, and economic capacities.

Related studies in the United States have indicated
that the enlargement of farm units may have detrimental effects on the quality of life in rural
communities, in terms of reduced services and
business activities in rural centers. (10) On the
other hand, capital-intensive production has greatly improved efficiency, allowing for "a more rapid
rate of growth in farm output on large farms while
small, marginal farms may be forced to leave the
industry." (11) Some social science researchers
maintain that large-scale farm operation induces
detrimental "stress" effects on farm families and
affects the health of the operator, who may be
short of the skill necessary for modern production
and marketing practices. Likewise, the education,
transportation, and service sectors have suffered
as the farm population has decreased.

In summary, the family-farm pattern may be
expected to remain as the crucial characteristic of
Canadian agriculture, but it is likely to adjust
its production and marketing activities to the
model of a modern, industrialized society. Government policies designed to assist small farmers and
to make their farms viable were established in the
early 1970s. The Small Farm Development Program
(SFDP) operates on a national level. Under this
program, two types of counseling services are
offered. The first is directed toward farmers who
are leaving the farm, either to retire or to pursue
other employment. The second is directed toward
farmers who wish to to improve their incomes on the
farms, with advice on how to improve the efficiency
and productivity of their operations. This program
is extremely important in enabling small farmers to
stay in agriculture and in providing for part-time
farmers, creating an opportunity to examine alternatives for gainful employment. (12)

THE CHANGING RURAL LANDSCAPE

The characteristic feature of modernization in
Canada is increasing urbanization. The rapid growth
of Canadian cities causes them to extend further
and further out into rural areas, increasingly
blurring the division between town and country.
This makes it very difficult to identify any
specific demarcation between rural and urban populations. Consequently, Statistics Canada has recently revised the definitions of "urban" and

"rural": "Urban is the population living in each
concentration of population of 1,000 inhabitants
and over, with a population density of 1,000
inhabitants and over per square mile. The remainder
is rural." Using this new typology, the rural farm
population represents 18.64 percent and the rural
nonfarm population represents 19.56 percent of the
total population. The remaining 61 percent is
urban. (13) The constant trend towards increased
urbanization has greatly affected the way of life
of the rural population and has defined their
aspirations both in economic and social matters.
This has manifested itself in changes in the
standard of living, which have occurred both in
urban and rural communities. (14)

The intent of Canadian agricultural policy is
either to preserve the existing rural/urban ratio
in the population or at least to slow down the rate
of increase in the urban population.

Depletion of the rural population will, in general, have a detrimental effect on the rural
community. Furthermore, many studies have shown
that rapid expansion of the largest urban
centers typically brings in its wake sharp
increases in problems of social maladjustment
with attendant social and economic costs. It
has been even suggested that the quality of the
city dweller's life is enhanced if he has a
populated countryside to visit from time to
time. For this reason, a strong positive relationship is shown between a balanced rural/ur-
ban population distribution and the national
goal of an improved quality of life. (15)

Attempts to improve the quality of life demand a
comprehensive approach. It is essential to examine
the impact of the threat of technological unemployment in the rural community, including its farm and
nonfarm components. The sector which has suffered
most from the modernization of farming is the rural
farm service sector. The fundamental transformation
and improvement of the transportation system since
the end of World War II has minimized the demand
for rural services. Rural people now purchase farm
input and household goods in large urban centers.
Railway lines have been abolished; most rural
families own automobiles, and bulk transportation
has been taken over by long-distance truck services
centered in the large cities. The decline in the

rural population has caused a reduction in the
number of schools and diminished the availibility
of medical services. In such circumstances, small
towns and the open country cannot hold people
unless specific measures are taken to improve the
economic environment. To maintain the size of the
rural population, it is therefore necessary to
design programs so that all sections of the rural
community benefit from it. In other words, rural
development policy has to coordinate available
resources to strengthen the social and economic
power of rural communities. This calls for the
coordination of agricultural policy program with
broader-based rural policy. A prime goal of such
coordination is to provide access to public services in nonmetropolitan areas. The sine qua non to
success is, however, local participation. This may
pave the way for integrated local development aimed
to achieve a modern, viable farm system linked with
a prosperous, rural nonfarm community. To achieve
this goal, several problems must be resolved. With
a great deal of simplification, these may be
summarized as the effects of farm population growth
generating migration, underemployment, unemployment, and poverty.

   Over the two decades from 1951-1971, the farm
population in Canada decreased by 49.8 percent
while the rural nonfarm population increased by
46.4 percent. (16) The growth of the rural nonfarm
population occurred mainly in the late 1960s and
has continued since. Most recently, a movement out
of the large cities has gained momentum. Thus, the
growth of the rural nonfarm population has been
nurtured by two sources: overpopulated farms and
metropolitan centers. The high rate of increase of
the rural nonfarm population is the prime cause of
the high rate of unemployment in the rural labor
force, as reflected in Appendix 1.3. These statistics also illustrate the changes that have occurred
in the economic structure of rural communities.
Existing village crafts are disappearing rapidly,
while a new pattern of demand is developing along
with the modernization of agriculture. One way of
meeting the damaging trend of increasing unemployment is the creation of industries adjusted to
the demand for new inputs and the manufacture of
consumer goods that could be produced locally. As
far as employment is concerned, these industries
might provide new outlets for local personnel. In
general, potential industries are small, and their

spatial distribution would be determined by local opportunities. This supports the argument for a "locally based development strategy," as advocated by the Council on Rural Development. Rural industries would surely increase the opportunities for unskilled, part-time farmers.

The pace of modernization in Canadian agriculture and the extent of social dislocation has generated a large body of literature on rural poverty. The data in Table 1.3 would suggest that almost one-fourth of Canadian farm families are poor and living below the poverty line. However, farm produce as a measure of farm income is a poor guide. But information on off-farm income, which probably radically alters the situation, is scarce.

The period of large-scale migration to the city has ended. Those farmers who remain now expect - indeed, demand - a better life than before. Between 1961 and 1971, the total number of farms in the six eastern Canadian provinces decreased by 31 percent. Over the same period, the number of low-income farms fell by 55 percent. This would suggest that, among other forces, farm adjustment programs designed in part to meet farmers' new objectives were effective in reducing the number of low-income farmers. (17)

INSTITUTIONAL PROGRAMS ON RURAL ADJUSTMENT

Major government intervention to foster the economic viability of rural communities and to promote their development dates back to the early 1960s. The prime constraint to successful performance of the early programs was the disagreement on authority. Under the British North America Act, provinces have jurisdiction over natural resources and draw up their own legislation to manage and develop the land base. Most importantly, then, federal legislation must rest on the close cooperation and agreement of provincial authorities. This is not always forthcoming. This split in jurisdiction and an ever-increasing drift toward greater provincial autonomy are fundamental weaknesses of much regional development legislation.

The Agriculture Rehabilitation and Development Act of 1961 (ARDA) was centered on farm problems. A major objective was first to raise incomes in rural areas, including those of farmers. The main thrust in this was to help farmers to step up into a higher economic echelon (see Table 1.3). As a second objective, emphasis was laid on measures to promote soil and water conservation, on land conversion programs, or, more generally, on better allocation of resources. This program also included the goal of farm consolidation. Confusion remained, however, as to whether the aim should be to increase the intensity of production to retain a large farm population, or whether young farmers, in particular, with small farms in marginal environmental areas should be persuaded to move into other sectors of the economy. Initially, at least, support was given to the retention of a large farm population. A major shortcoming of the program has been the inefficiency of extension (training) services. Their success has been limited to progressive farmers with the propensity to adopt technical innovations. However, the marginally educated farmer has been generally neglected. The lack of dynamic personnel, the limitation of funds, and the refusal of some farmers to accept instructions from "strangers" have all contributed to the lack of efficiency in extension services.

In the case of ARDA, funding arrangements were cost shared 50-50 between the provinces and the federal government. In the first five years of its existence, the commitment of federal funds under ARDA amounted to $65 million, of which about half has been spent. No more than token efforts characterized ARDA actions. (18) It took four to five years to recognize that inherent problems in rural areas are complex and that measures must be focused not only on agriculture but on the nonfarm sector as well. The Canadian Land Inventory, established in 1963, was maintained to provide a wide-ranging assessment of land capacity, including its potential for agriculture, forestry, recreation, and wildlife, and embracing the extensive range of social and economic factors directly relevant to land use. This was seen as a vital prerequisite to effective rural development.

Subsequently, a number of other pieces of legislation were introduced which together form the basis for Canada's current rural-development In 1966 a special fund, the Fund for Rural Economic Development (FRED), was established, having a wider range of activities than previous programs. FRED was designed to promote social and economic development in specially designated development areas,

to increase income levels and employment opportunities, and to raise living standards in those areas.
The program included schemes for education, infrastructure, services, and rural industrial development, including tourism. It also monitored housing,
recreational parks, and local participation. Indeed, the various FRED programs assigned a high
priority to local participation, but often failed
to win full acceptance in the communities involved.
Since the FRED programs have been concentrated or
even limited to "pockets of poverty," most of the
funding originated from provincial government
funds, and financial support from the federal level
has gradually decreased. A special study of the
Economic Council of Canada summarized the weakness
of those schemes, stating that the federal government

> failed to coordinate these programs and poli-
> cies towards any specific goals for any region.
> At the same time, the individual provinces also
> instituted regional development policies of
> their own, with the result that there was a
> hodgepodge of development efforts that lacked
> any general overall direction. (19)

To eliminate the conflicts in goal setting and
regional discrepancies in growth, the federal government established the Department of Regional
Economic Expansion (DREE) in 1969.

In 1973, DREE completed an extensive policy
review, which concluded that existing programs were
often too limited in scope and tended to be
oriented too much to local problems rather than to
the potential of an area. The revised policy
required the negotiation with each province of a
ten-year General Development Agreement (GDA), which
would not identify the specific regional programs
to be followed, but which would rather provide a
framework within which special agreements could be
drawn up to meet local opportunities. Direct assistance to and the active participation of the
private sector is an important feature approach.
Over the period 1969-1975, assistance to the private sector accounted for 30.7 percent of total
DREE expenditures. Nonrural infrastructure (highways, streets, and water and sewer systems) accounted for 35.4 percent, while rural assistance
accounted for 29.3 percent. In total, $1,551 mil-
lion was spent. (20) The policy review also ac-

cepted that each province is unique in its economic
and social composition. With this in mind, provincial authority has been further strengthened, and
increased attention is now also paid to continuing
provincial and federal consultation on development
issues. A wide range of different development
projects is anticipated, with a number of different
federal-provincial cost-sharing schemes, depending
on the financial means of the province concerned.

The Canadian Council on Rural Development
(CCRD) has been an entirely nongovernmental organization in its composition. It evolved out of ARDA
and a number of credit and farm counseling services
under the federal Department of Agriculture. It
operated as an advisor to DREE, particularly in
problems of rural unemployment and underemployment.

The CCRD advocated a "locally-based development
strategy" with emphasis on community involvement
and a suitable scale of rural development. One of
the prime tasks of CCRD was to provide more
effective means for adjusting the conflict and
unjustices that invariably arise as technological
and other changes take place.

Two other aspects of modernization have
sparked new legislative interest. First, the national phenomenon of land transfer from agriculture
to other use has been mentioned in an earlier
section of this paper. Nearly one-third of Canada's
population is now concentrated in three metropolitan regions (Montreal, Toronto, and Vancouver), and
there is a constant growth in the densely populated
industrial belt that runs from Montreal to Windsor.

Gradually, one province after the other has
established zoning laws that focus on the preservation of farm land areas to sustain the balance of
nature, including parks, recreation centers, and
the like.

Second, the modernization of agriculture has
increased the demand for energy. Attempts have been
made to utilize locally available and renewable
resources for both farming and rural holds. Furthermore, modernization has put into the fore-
ground the relationship between agriculture, rural
communities, and ecological considerations. It is
essential that modern agriculture operates in an
environment that preserves the quality of life for
the entire rural population. The "locally based
development" may be an avenue for establishing, on
the one hand, a greatly desired economic equilibrium and, on the other, social harmony.

ASSESSING THE FUTURE

In the years ahead we can expect to see a continuation of the trend toward modernization in Canadian agriculture. The potential for technological progress in farming is well recognized. At the same time, urbanization is likely to expand. However, a growing number of constraints (transportation, pollution, housing costs, and others) will limit continued urban growth. Under continued pressure from agriculture and increased urbanization, the rural nonfarm population will be subject to a new policy focus in which the major objectives will be: to maintain the ecological equilibrium; to promote local industries producing agricultural input and output items; to reorganize the structures of the service sector; and to assist the growth of smaller communities and the development of new communities. But future public policy must also encourage part-time farming, the introduction of small-scale industries, perhaps a new kind of cooperative movement, and the establishment of a network of extension services serving both the rural-farm and the rural nonfarm population.

NOTES

(1) From 1971 to 1976 total improved land increased by 5 million hectares and the area under crops by 3 million hectares.

(2) Agriculture Canada, Orientation of Canadian Agriculture, Vol.1/A, 1977, p. 42.

(3) Ibid., p. 54.

(4) For example, it was reported in November 1973 that in the previous 12 months, 7,356 dairy farms in Quebec disappeared. See D.L. MacFarlane and L.A. Fischer, Canadian Dairy Industry, Food Prices Review Board, Ottawa, 1974, p. 22.

(5) W.D. Jones, "Farm Labour, Agriculture Canada," Market Commentary, December 1978, pp. 69-82.

(6) Ibid., p. 69. In 1976 less than 35 percent of all farms used hired labor and less than 7 percent employed paid workers on a yearly basis.

(7) In 1975 the total investment in machinery and equipment accounted for $6.4 billion.

(8) Canada Grains Council, Statistical Handbook, 1974; Canada Agriculture, Market Commentary, 1980.

(9) The highest share (70 percent) of farm income is in Saskatchewan and the lowest (12 percent) is in British Columbia.

(10) Small Business and the Community, A Study in Central Valley of California on Effects of Scale of Farm Operations, U.S. Government Printing Office, Washington, 1946.

(11) Jones, W.D. and Fui-Lai-Tung, "A Regional Comparison of Structural Change and Resource Use in the Canadian Farm Industry 1961 to 1971," Canadian Farm Economics, Vol.5 (1977) 12. pp. 20-30, Ottawa.

(12) Canadian Council on Rural Development, Job Creation in Rural Canada, Ottawa. 1978, p. 32.

(13) T.F. Parenteau, "Typology of the Canadian Population," (Working Paper) No.4-GED 78, Statistics Canada, 1978) pp. 8, 14, 68.

(14) R. Shaffner reports (Observation No.19 of the C.D. Howe Research Institute, 1979) that the percentage of the total Canadian housing stock with piped hot and cold water increased from 56.9 in 1951 to 97.8 in 1977; the share with flush or chemical toilets rose from 70.7 to 98.3 percent over the same period; and the share with installed bath or shower facilities rose from 60.8 to 97.8 percent.

(15) Agriculture Canada, Orientation of Canadian Agriculture, Vol. II (1977), p. 19.

(16) Canadian Council of Rural Development, Job Creation in Rural Canada, (Ottawa, 1978), p. 70.

(17) G. Steele, "The Canadian Experience in Rural Development," in Managing Rural Development (Macdonald Stewart Institute of Agriculture, Pointe Claire, Que: 1978), p. 32.

(18) H. Buckley and E. Tihanyi, Canadian Policies for Rural Adjustment: A Study of ARDA, PFRA - and MMRA (Ottawa: Economic Council of Canada, 1967), p. 15.

(19) Economic Council of Canada, Living Together: A Study of Regional Disparaties, (Ottawa, 1977), p. 147.

(20) Ibid., p. 151.

Appendix 1.1. Land With Some Capability For Agriculture Classified by Soil Capability By Soil Capability Class And Climate Zone, Canada, 1971 (in thousands of hectares)

Soil Capability Class	Climate Zone (ACRI)*						
	A	B	C	D	E	F	Total
1	610	851	852	1,696	325	0	4,334
2	329	1,513	1,304	9,046	4,739	59	16,990
3	221	817	1,632	10,452	12,509	300	25,931
4	14	336	2,537	9,431	9,863	2,873	25,054
5	4	199	1,247	7,510	13,028	5,392	27,380
6	33	265	790	3,882	7,719	1,436	14,125
Total	1,211	3,981	8,362	42,017	48,183	10,060	113,814

Sources: (1) Canada Land Inventory Soil Capability
(2) Williams, G.D.V., Population, Technology and Resources, Report 25, Science Council of Canada, 1976.

* ACRI – Agroclimatic Resource Index which rates the climate of Canada's agricultural land by frost-free season, heat accumulation, and moisture deficiencies.

Reproduced From Agriculture Canada, Orientation of Canadian Agriculture, Vol. 1/A (1977), p. 37.

Appendix 1.2 Average Annual Employment in Canadian Agriculture by Class of Workers 1966-70, 1971-75, 1976, 1977 and 1978 (in thousands and percent of total[a]

Year	Total	Paid Workers		Unpaid Family Workers		Self-Employed Operators	
1966-70	529	98	18.2	120	22.3	321	59.5
1971-75	482	101	21.0	106	22.0	275	57.0
1976	474	143	30.2	90	19.0	241	50.8
1977	468	145	31.0	84	7.9	239	51.1
1978[b]	473	130	27.5	92	19.5	251	53.1

[a]Data might not total because of rounding.

[b]Unofficial estimate.

Note: Data for 1976, 1977, and 1978 are based on the revised labor Force Survey and are not comparable with the earlier data, especially by class of worker.

Source: Statistics Canada, The Labor Force, Catalogue No. 71-001, Monthly.

Reproduced from W. D. Jones, "Farm Labor,: Market Commentary, Ottawa: Agriculture Canada, 1978).

Appendix 1.3. Rural Labor Force, Employment, and Percentage Unemployment for Canada, 1976.

	Labor Force (thousands)	Employment (thousands)	Unemployment (percent)
Total	1,733	1,589	8.3
Agriculture	405	397	1.9
Non-agriculture;	1,320	1,193	9.6
Primary industries	107	94	12.1
Manufacturing	266	238	10.5
Construction	146	124	15.1
Transport, Communication	127	118	7.1
Trade	217	202	6.9
Finance, Insurance and Real Estate	36	34	5.5
Service	336	304	9.5
Public Administration	85	78	9.4
Unclassified	8	-	-

Source: Canadian Council on Rural Development. Job Creation in Rural Canada (Ottawa, 1978), p. 64.

2 Agriculture and Rural Change: The Case of France

Jean-Claude Bontron
Nicole Mathieu

A great deal has been written about the changes
which have taken place in both agriculture and the
rural areas of France since 1950. In regard to
agriculture, the general tendency has been for a
spectacular increase in the absolute quantity of
production and, most especially, in land and labor
productivity. This has been accompanied, however,
by a lowering of the contribution of agriculture
both in terms of the domestic product and the
national value added total, and, moreover, by an
accelerated decline in the number of persons employed. Although they are only exceptionally and
locally competitive with other forms of uses with
which they are in competition, agricultural activities have maintained their teritorial control
through the appropriation of more and more land.
As for rural space over the same period of time,
new tendencies have emerged, placing more and more
demand on the acquisition of land. In addition,
this space, because it is increasingly empty and
therefore more open to other forms of use, has
become an object of multiple conflicts. Rural
areas today are sought after for the creation of
new profits through industrial redeployment and the
emergence of mass coastal and mountain forms of
tourism, for the realization of those facilities
that demand large amounts of land (transportation
systems, energy production, and public services).
Finally, in reaction to increased urbanization of
the consumer society, rural space is also sought
after as a refuge by various marginal groups
(retired people, neo-ruralists, the sick, the in-

firm, and various sects), and for the preservation
of scarce resources and threatened ecosystems. Less
and less agricultural in terms of employment, the
countryside is no longer the same labor reservoir
it was in the nineteenth century, but tends to have
become a reserve of space. Changes in terminology
and the replacement of the term "rural space" for
"the countryside" captures this idea well.

But if these agricultural and rural changes
are well known in broad outline, the links between
changes in agriculture and changes in rural space
have been appreciated to a much lesser extent. It
seems clear that the interactions and interde-
pendancies between an economic sector and spatial
processes have not been clearly identified. Our
intention in this chapter is to examine the reasons
for these omissions and to propose some answers
from work that has been carried out on the relationship between agricultural and rural change.
The limited state of knowledge in regard to this
question would seem to arise from a growing dichotomy between agricultural analysis and rural analysis in the spheres of research, applied science
and planning, and forecasting. In rural research
organizations, in geographical, sociological, and
rural-economic research groups, one can recognize,
in regard to the major changes that have affected
agriculture and rural areas, a degree of specialization in each of the two branches of the dichotomy. The consequence has undoubtedly been an increase in knowledge levels, but along parallel
paths, without any attempt at the much-needed
linkage between the two.

Thus, the interpretation of the changes which
characterize French agriculture have given rise to
numerous conflicting theories about the processes
involved. The thesis relating to the development
of agrarian capitalism has been abandoned, given
the decline in the salaried labor force, the weak
division of labor in agriculture despite technical
progress, and the weak development of large holdings. Other theses have therefore been advanced.
One is that of the proletariatization, in atypical
form, of agriculturalists, who have tended to
become domestic workers for the food processing
industry; another is that of the reinforcement of
small-scale commercial production, as desired by
capitalism, to such an extent that the nonindustri-
alizable character of the agricultural work process

and the land problem oppose the intrusion of
capitalist forms of relationship. Finally, there
is the thesis of the demise of the peasantry and
its reconstitution into new social strata, with the
development of localized forms of agrarian capitalism. All these theories have advanced our knowledge of the social differentiations, which agricultural modernization or the penetration of capitalism into the agricultural realm have introduced
into France, by recognizing categories of agriculturalists whose behavior and future prospects were,
in the last resort, little known (for example, poor
retired peasants, those with several occupations,
and integrated livestock farmers). In the same
way, understanding of regional variations resulting
from agricultural change has also advanced. In
particular, the reinforcement of regional specialization has been highlighted by the sharp divisions
that exist between the cereal-growing areas of the
Paris Basin and the regions of livestock production
in the west and also divisions in descriptions of
movements in the region of pig production and its
extremely strong concentration in Brittany. The
critical consequences of the common agricultural
policy and the implications for such specialized
Mediterranean regions as the winegrowing areas of
Languedoc, have also been analyzed. Finally, regional systems and types of agriculture, which have
resulted from these changes over the last thirty
years, have been particularly well studied. One
could extend the list further, concerning this type
of research, but the important point is that these
results have rarely been considered in relation to
those other changes that have been taking place in
rural space.

In regard to the other sector of research, it
is pertinent to recall that the first theoretical
contribution of pioneer analysts of the rural scene
was the affirmation of the need to draw a distinction between the rural and the agricultural, and
the refusal to consider the evolution of agriculture and the evolution of the French countryside in
an integrated manner. The main contributions in
this realm have been the recognition of the differentiations that have occurred as a result of
urbanization and urban out-migration (at retirement), and redeployment in tourism and industry.
Studies of declining and aging populations have
often been considered along with the degradation of
services ("desertification"). Although the theo-

retical character is less well developed (for the
spatial analytic tradition is not as old), contro-
versies are still to be encountered, for example,
in identifying the place of rural space in postwar
society (a space for profit or a space to be
abandoned): in defending or, on the contrary,
contesting the notion of "production space"; and,
most particularly, in developing typologies of
rural space. In regard to this latter point,
consensus ... far from having been achieved, and
research groups continue to debate the question of
the number of space categories, their designations
(for example, periurban or urban, low density,
marginalized, or devitalized), and most of all the
methods to be used in identifying them. Should the
approach be quantitative and based on a series of
criteria, some of which are spatial in character
(for example, periurban zones of low density), or
should the stance be theoretical and based on the
struggles and strategies of capitalism and the
degree of integration of different space types?
Whatever the contradictions and debates, they have
led to a clearer appreciation of the major changes
that have taken place in rural space, so much so
that it has led to an improvement in the quality of
state intervention in town and country planning.
In following this policy from the "pilot sectors"
to the system of regional contracts, one has
learned to recognize the agents that lie behind
these transformations, (property promoters, indus-
trialists, local authorities, and various agencies
of the state), and the forms of resistance to this
intervention (such as land struggles and ecology
movements).

However, in this endeavor to analyze the
mechanisms of change, the place of agricultural
transformation, based on the types of agriculture
and on the different social layers within the
peasantry, is only rarely dealt with precisely.
Even in rural geography, which constitutes in
principle a complete systematic branch in its own
right and where, in regional analyses in particular, all aspects of change are broached, one can
often recognize the juxtaposition of agricultural
and rural approaches, even when this is not a
specialty of the researchers in either of the two
domains.

Even more serious in our view is the fact that
this dichotomy also touches the applied sciences
and administrative agencies. Here we will give

just two examples to illustrate this contention. The first is that of urbanism, an applied science which has largely evolved since the war in parallel with the process of urban growth. The urbanists, who do not exactly constitute a distinct body in their own right, come from diverse backgrounds (the great engineering schools, schools of architecture, and university departments of urbanism and urban economy), but they have never been trained in matters relating to the agricultural industry and peasant society. However, in practice, a certain number of them cannot avoid encountering such issues. Thus, in the team of urban planners associated with the new town of Sergy Pontoise, there are no agronomists, but there are young, enthusiastic engineers, architects, sociologists, and landscape architects attached to the idea of living on the site, all imbued at the outset by the idea of free space and conscious of the task of implanting a new town in the country. They know nothing of agricultural production systems, of landholding systems, of agricultural groups, before they prepare land utilization maps within the planning area of the new town.

The second example of this disassociation in planning practice is seen in the evolution of the role of the ministry of agriculture in rural planning. Faced with the changes which have taken place in the French countryside, where since 1962, agricultural workers are no longer in the majority, the ministry of agriculture has adapted its action to the facts of the new situation. While remaining the adviser on agricultural questions, the ministry has tended to enlarge its prerogatives in order to maintain a certain degree of power in rural areas and, thereby, to control the interactions between the agricultural sector and exogenous forces, and a rural planning service has even been created, becoming a distinctive institution in its own right in 1965. This institution has been progressively endowed with research organizations, first at a national level, then at the departmental and regional levels. This institution has conceived and put into operation its own intervention procedures, in particular the Rural Management Plans (PAR). But the recent tendency is undeniably the expansion of the responsibilities of other ministries, while the ministry of agriculture itself has become more and more restricted in its role as a ministry for agricultural production. The new area planning policies which effect rural space (less favored farming areas, policies for the mountain regions, regional contracts, national and regional parks) do not depend on the ministry of agriculture and, moreover, are no longer in the last analysis agricultural, as around the year 1970. The policy for mountain regions has progressively passed from one objective: the preservation of an agriculture threatened by the demands of modernization to that of the development of new economic sectors and the priorities of conservation. Agricultural policy, therefore, is more and more dissociated from town and country planning, which continues to rest on a partial understanding of the agricultural world and on a unique conception of agriculture: that of the agriculturalist as a guardian of ecosystems and the landscape, who holds several jobs and welcomes tourists, and who tends to occupy a position of low skill in a factory in the countryside.

It is obvious that we could have found counter examples to this general tendency of growing dissociation between agricultural and rural analyses, but it seems possible to us to subscribe to the view that this tendency is of major import. In an applied context, this tendency expresses itself in the conflicts between various pressure groups, which lead either to an overestimation or underestimation of the significance of agriculture and agriculturalists according to the standpoint adopted. In a research context, it expresses itself by the real theoretical difficulties posed by the linkages and discontinuities between the two fields of inquiry. The relationship between agricultural activity and spatial processes, between economic analysis and spatial analysis, poses a key theoretical problem for which hypotheses are lacking.

If it is not possible to find a clear knowledge of the links between rural and agricultural transformation within the research and rural planning practice, one may try to analyze the facts using a statistical analysis of available data. The analysis of changes in agriculture and in rural areas using statistical methods certainly does not describe the complete range of diverse contradictions existing in these areas. Nor does it define with certainty the complex range of different internal and external pressures which determine these changes. It is, however, necessary to refer to this analysis because it facilitates the precision of interrelations between different indicators of change and consequently shows tendencies toward autonomy or integration of the agricultural or rural factors in the changes of these areas.

Our results are based on a research program carried out over many years, looking at the changes and typology of rural areas, and based on a systematic use of the statistical evidence available for these areas. Initial analysis based on the departmental level (1) was followed by a more refined approach based on a sample of 220 "France" representative "cantons". (2) The canton is a particularly suitable unit for the analysis of rural areas transformation because of its metric characteristics, its roots in the past, its important role in the social organization in rural society, and, above all, in the nature of the available statistical data. This "data bank" of rural changes enables the statistical approach to present an important tool in the analysis of changes in the countryside. (3)

An initial set of questions may be asked about the role of agriculture in employment and rural settlement and, above all, in the relationships between the decline in the number of agriculturalists and the demographic and economic evolution of rural areas.

In 1954 agriculture directly employed 57.5 percent of the active rural population. When the services allied to agricultural production and population are included, this explains the employment of 75 percent of the rural population. One notes, therefore, that there exists a significant correlation (.79) between the density of active agriculturalist males and the total population density in rural areas. The rural regions with a high density of agriculturalists have a high global population density, which does not preclude the inverse (-.26) between the percent of agriculturalists and the density level. At this time, variations in farm-working settlement levels are to a great extent explained by those of rural settlements.

Between 1954 and 1975, considerable changes affected the economic basis and settlement pattern of rural areas. The active agricultural population dropped 60 percent with an incessant annual rate of decrease, such that the active agricultural population represented over 32 percent of the total rural population at the end of the period. The absolute and relative number of nonagricultural population

increased mainly because of the steady increase of
commuting to urban areas. The active rural population working in towns exceeds one million in 1975,
which explains nearly 20 percent of the rural
settlement. The commuter number is relatively constant in spite of the regular expansion of the
"urban areas" which indicates that the commuting
radius enlarges more and more (see Table 2.1). At
the same time, nonagricultural activities located
in rural areas will maintain their potential of
jobs because of a progression in the tertiary
sector heavily characterized by the predominance of
declining traditional particular industries (such
as mining, leather textiles, and forestry).

Table 2.1. EVOLUTION IN RURAL EMPLOYMENT

	Employment	Percent 1962	1975	Employment/km2 1962	1975
Primary	1,704,735	56.4	40.4	7.02	3.65
Housing	418,155	6.8	9.9	0.85	0.89
Industries	868,055	16.1	20.6	2.00	1.86
Agricultural	156,800	2.7	3.7	0.34	0.34
Tertiary	1,225,675	20.6	29.1	2.56	2.62
TOTAL	4,216,620	100	100	12.44	9.02

Finally, in spite of the collapse of rural
employment between 1954 and 1975, rural population
decrease is only 6 percent within the same rural
limits, with a geographical reallocation that is
clearly illustrated by Figure 2.1 (representing the
period 1962-1975).
While certain regions emerge with an accelerated demographic decline (in particular, the Massif
Central and its boundaries), others are able to
stabilize and even at times to steadily increase
their rural population.
The Table 2.2 intercorrelations between the
different trends of the diverse population categories clarify the relations between agricultural
evolution and rural evolution.
One may note that in spite of its important
influence at the beginning of the period and the
amplitude of the trend, the decline in agricultural
population is not significantly correlated either
with changes in active nonfarming population or
with the variations of total rural population

areas.

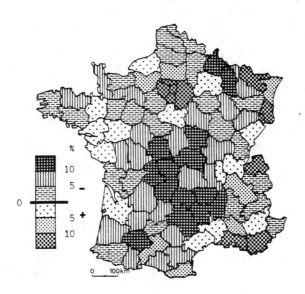

Figure 2.1. Rural Population Evolution 1962-1975.
Basic unit: "cantons a dominante rurale" (rural
population of the canton is greater than urban
population).

Table 2.2. Intercorrelation Between the 1954-1975
   Trends Affecting Different Strata of Populations

1954-1975 variation population	(1)	(2)	(3)	(4)	(5)	(6)
1. Total active	1					
2. Agricultural active	0.21	1				
3. Nonagricultural active	0.80	0.01	1			
4. Total	0.94	0.05	0.86	1		
5. Rural	0.69	0.18	0.69	0.75	1	
6. Urban	0.71	0.02	0.76	0.83	0.52	1

Evidently, the "cantons" with the highest
agricultural level at the beginning are precisely
those which submitted the strongest depopulation
(.43 correlation between those two variables), but

here the influence of past structure, not recent transformations, in agriculture can be seen.

Newly created local nonagricultural employment did not, contrary to planners' opinion, play a determinant role in the active agricultural population decrease. However, it becomes fundamental when explaining transformations of the rural economical base and of the demographical evolution. More precise analysis lead us to discover that it is the creation of tertiary employment, and particularly female employment, that was the most determining factor for stabilization or growth in rural areas. It is very interesting to observe that, concerning the cantons, the rural population evolution is strictly correlated with that of the urban population while it is not correlated with the agricultural population trend.

External factors to agriculture and to "rural space" (for instance, the phenomenon of exurbanization) are determining the rural transformation process. Hence, the tendency is toward stronger autonomy in rural demographical evolution vis a vis consequences of modernization upon agricultural employment.

The last question is: What is the relation between the land market changes and agricultural and nonagricultural land demand in rural areas? Statistical data are much less complete on this matter. However, certain facts can be outlined.

In all of the French regions, land prices increase steadily (more than the inflation rate) despite their agricultural evolution and, in particular, despite the obvious decrease in farming land in geographically less favored areas. An increasing nonagricultural land demand takes place in all different types of rural areas--of course those situated on the outskirts of urban areas and on the coast, but also in deserted areas which are today the object of multiple conflicts concerning new private or state land uses (such as reforestation, nature preservation, military areas, and dams). This nonagricultural land demand, insignificant in surface quantity (0.15 percent of cultivated surfaces per year) is dispersed: thus it provokes speculative anticipations which determine, in all places and for every kind of transaction, a land price increase.

Therefore, the external factors are much more important in the evolution of the land market than the transformations in the agricultural sector, despite the heavy competition between farmers. This is why the Table 2.3 correlations do not bring out any link between land price evolution and agricultural factors.

Table 2.3. Intercorrelations Between Some Criteria of Land Use.

Variables	(1)	(2)	(3)	(4)	(5)
1 Arable land price 1965	1				
2 Price increase (1965-1971)	-0.44	1			
3 Cultivated land variation (1963-1970)	-0.06	0.20	1		
4 Urbanization level	0.57	-0.09	-0.11	1	
5 Rate of new inhabitants in rural communes	0.53	-0.15	-0.08	-0.52	1

These results confirm the preceeding observations. In any case, they prove that the phenomenon external to agriculture prevails over the rural transformation processes. On one hand, one notices an autonomization of rural change vis a vis all agricultural change; on the other hand, nonagricultural events do not seem to disturb a certain logic in the specific evolution of agriculture.

NOTES

1. See J.C. BONTRON, M. SACHET, and L. VELARD, L'espace rural francais, statistiques et indicateurs, (Paris: ACEAR, 1972).

2. The canton is an administrative district; in rural areas the average characteristics are 12 "communes" for 1 canton, 200 square kilometers in area, with 6,500 inhabitants.

3. More than 200 variables available.

3 Socioeconomic Effects of Agricultural Development and Rural Transformation in the GDR

W. Roubitschek

Agricultural policy in the German Democratic Republic is directed towards two main, closely linked goals: to increase production and its efficiency in agriculture and the food industry, in order to guarantee a stable, constantly rising supply of high-quality food to the population and of raw materials to industry; and to bring living conditions in the countryside closer to those in urban areas, thus gradually overcoming the existing differences between town and country." (1)

Since the population's demand for food and the industry's demand for agricultural raw materials are continuing to grow both quantitatively and qualitatively, agricultural production must also continue to grow. Other reasons which have to be taken into consideration by the socialist states are the growing world population and the international situation with regard to such matters as raw materials and price developments. This increase can, among other things, only be achieved by the creation of large-scale units and further specialization and concentration of production. According to the decisions of the Eighth and Ninth Party Congresses of the Socialist Unity Party, the future development of socialist agriculture is marked by intensification and the gradual transition to industrial production methods through greater cooperation.

Table 3.1 characterizes the GDR's agriculture in 1976. There are just under 6.3 million hectares of agricultural land for the supply of food for the GDR's population of about 16.8 million; that is,

0.37 hectares per head. This is the smallest agricultural area in all the socialist countries, with
the exception of Albania. Farming land cannot be
extended; on the contrary, agricultural areas have
to be given up because of the necessity to extend
brown soft coal mining and the building of houses,
roads, and industrial plants. Thus, there is only
one way open for meeting the rising demand for
food: further intensification of production; that
is, the more efficient use of labor on the same
acreage. The main factors of intensification are
the complex utilization of the results of science
and technology and the better control of scientific,
technological, and socioeconomic processes in their
unity.

The primary directions of intensification in
agriculture are toward the use of chemicals, mechanization, land improvement, technological drying,
progress in plant cultivation and stock breeding,
and training of agricultural staff. The growing use
of inorganic and organic fertilizers, herbicides,
and pesticides has a significant share in the
increase of yields (see table 3.1). In 1977, the
turnover of fertilizers in agrochemical centers
(ACZ) was 7.7 million tons. Plant protection was
carried out on 7.8 million hectares and 4.2 million
hectares were cultivated from planes (fertilization,
pest control, and, in some cases, even sowing of
grain).

Mechanization aims, above all, at the replacement of human labor by machinery and at making work
easier. Complete sets of machines are characteristic of industrial production. The proportion of
combines and other highly specialized machinery as
well as of fast-moving tractors and trucks has
rapidly increased.

Important factors in intensification are: an
effective arrangement of fields, which meets the
demands of modern productive forces (fields of more
than 80 hectares); the tendency toward homogeniza-
tion of soil conditions; irrigation and drainage;
and influencing and controlling natural site conditions and natural cycles as a precondition for the
achievement of high yields. In recent years, over-
head irrigation of sites with low ground water
levels in dry periods has gained more and more
importance. Between 1970 and 1978, the production
of dry fodder in the GDR increased from 597 kilotons
to 2,668 kilotons. More attention, however, must be
paid to energy saving methods. Moreover, while

considerable advances have been made in the training
of female staff, the percentage of full-time agricultural workers with complete professional or vocational training was only 5.9 percent in 1960, as
compared with 85.4 percent in 1978.

Another aspect which makes further increase of
production and productivity by industrial methods
necessary is the limited resources available to
agriculture. Characteristics of industrial production methods are, above all, the continued replacement of manual labor by machines and the carefully
planned concentration and specialization by means of
cooperation. Table 3.1 also shows the considerable
increase in productivity in grain production.

Table 3.1. Development of Grain Production
and Labor Productivity in the GDR

	Individual farming in the 1950s	Co-ops in the 1960s (Type III)	Agricultural production on a production line basis in the 1970s
Concentration of grain production in hectares per farm enterprise	8.0	320.0	2,500.00
Yield in quintals per hectare	22.0	26.0	40.00
Expenditure of labor in man-hours per quintal of grain	6.5	2.0	0.35
Gross production in 1000 Mark per man-power	13.0	42.0	240.00

While 15,711 state farms and cooperatives
worked an average of 374 hectares per enterprise in
1965, producing a multitude of different crops, at
the beginning of the 1970s a new set of decisions by
the Eighth Party Congress started a division of

labor into crop production and stock breeding and fattening. At the same time, the transition to industrial methods of production, mainly in crop production, was a process of specialization and concentration, as well as of cooperation between farm enterprises. On Februrary 28, 1979, there were in the GDR:

- 939 agricultural cooperatives (crop production)

- 23 agricultural cooperatives (horticulture, fruit and vegetable production)

- 62 state farms (crop production)

- 62 state farms (fruit and vegetable production, ornamental plants and tree nurseries)

- 166 cooperation communities and intercooperative establishments (crop production).

These enterprises employed an average of 280 co-op farmers and agricultural workers cultivating an average of 5,000 hectares of agricultural acreage with productive funds of nearly ten million marks. These agricultural enterprises specializing in crop production cooperate with 256 agro-chemical centers, 182 co-ops in soil improvement, 156 local agricultural engineering enterprises, and 247 building firms. 208 drying and pelleting installations were used for processing plant products. In addition, there were 104 plants for mixed fodder and 14 national enterprises for agricultural engineering.

Stock breeding and fattening is concentrated in 2,887 co-ops, 333 state farms, 30 national combines for industrial fattening, and 318 intercooperative establishments, with an average stock of over 1,500 gross animal units (1 gross animal unit equals 500 kilograms live weight).

Inevitably, the transition to industrial production methods in agriculture is linked with high investments. With basic funds per employee of 58,000 marks, agriculture has reached about 70 percent of the sum needed for an employee in industry.

The establishment of farm enterprises specializing in crop production and stock breeding is closely linked with the deepening of effective cooperative ties. Horizontal cooperation means cooperation with similar farm enterprises in crop production, the agro-chemical centers, local agricultural engineering enterprises, and enterprises specializing in stock breeding, in order to ensure such matters as the efficient use of machinery and the optimal supply of fodder. An expression of the close cooperation of farm enterprises is the rapid development of intercooperative establishments, mainly in building, soil improvement, and fodder drying; there were 1,482 in 1977, with 233,562 full-time employees. The coordination of targets is in the hands of councils of cooperation.

In addition, there is vertical cooperation in the form of associations; that is, cooperation between cooperatives and state farms, on the one hand, and processing industries and trade organizations, on the other. In grain, sugar, potato, fruit, and vegetable production, as well as in milk and meat production, more than four hundred cooperative associations were formed.

Finally, socialist economic integration, the close cooperation with the Soviet Union and the other socialist states, is increasingly affecting and penetrating agricultural living and working conditions.

Modern socialist agriculture is further characterized by close relations with the other branches of the economy (for example, the supply of vehicles, agrochemicals, and water). On the other hand, certain technological processes such as fertilizing, building, and transport are separated from the cooperatives and state farms. Progressively fewer agricultural products reach the consumer in an unchanged form; more than 85 percent of the production in agriculture is processed and treated in industry. While agriculture receives the means of production - raw and auxillary materials - from 115 groups of products, agricultural raw materials valued at about 700 million marks are used in many products outside the food industry (above all, in the light and chemical industries).

These expanding relations are pointing to the formation of an economic agro-industrial complex, which integrates all branches of the economy that are involved in producing food and agricultural raw materials. Apart from agricultural production itself, the spheres preceding it (such as branches of industry and services producing, maintaining, and repairing machinery and equipment) and the spheres following it (such as storing, processing, and trading agricultural products) are united in one complex system.

The total output of the agro-industrial complex, 119.2 thousand million marks, accounts for 30.6 percent of the total GNP of the GDR. 1.7 million employees, more than a quarter of all employees in the productive sphere, are involved in the production and supply of foodstuffs. One employee of the economic agro-industrial complex provides about ten inhabitants with food. Agricultural products have a share of more than 40 percent in the total turnover by the population in terms of value.

In the following paragraphs, some socioeconomic effects of the development of the GDR's agriculture will be discussed. In addition to the development of the socialist productions relationships and the ownership relationships in agriculture, as a consequence of the land reform and the formation of agricultural cooperatives up to the spring of 1960, and the parallel development of the class of cooperative farmers, it is above all the improvement of the farmers' working and living conditions that must be noted.

First, the territorial links between the structure of a settlement and the quota of the farm population in the total of the working population (agrarian quota) must be mentioned. On January 1, 1971, there were 17,068,000 inhabitants in the 8,868 urban and rural communities. If we classify the communities according to the proportion of those working in industry, agriculture, and the tertiary sector, we get the following scheme: Comparatively few communities, but most towns and cities, belong to the type of settlement dominated by the tertiary sector (more than 53 percent of all inhabitants). Over 36 percent of all communities, with just under six million inhabitants, live in settlements dominated by industry. More than two million inhabitants live in the 52.6 percent of all communities, which are dominated by agriculture. The majority of the 7,750 rural communities that have under two thousand inhabitants are oriented towards agriculture. They are inhabited by 26.2 percent of the population.

However, it is not possible to classify all communities with a small number of inhabitants as being dominated by agriculture. There exists a clear binary division in the regional distribution of rural communities. To the south of a line running along the Mittellandkanal and the central Elbe, in an east-west direction through the county of

Cottbus, there is a different basic structure, apart from the higher density of population and a more pronounced change from community to community. Mainly in the areas of industrial concentration (for example, around Karl-Marx-Stadt and Zwichau, in the Thuringian Forest, and in Upper Lusatia) the villages are, as a rule, inhabited more by industrial than by agricultural workers. In the northern part of the GDR, however, the towns are supply centers for a still predominantly agrarian environment.

This marked difference in the settlement structure between north and south in the GDR has a tremendous influence on socioeconomic development. Much more attention must be paid to the further industrialization of the northern areas, since there are regional reserves in the utilization of resources. Moreover, redirection of labor to southern areas can be avoided in the future.

In a number of studies, the changes in the housing conditions of the rural population are being investigated. In 1961 only 8.1 percent of homes in communities with under two thousand inhabitants were equipped with a bathroom or shower; in 1971 the proportion was 25.5 percent. The proportion of homes with inside toilet facilities rose from 9.1 percent to 18.0 percent. There was a particularly large increase in the connection to the public water mains, from 24 percent to 63 percent. Between 1945 and 1970, 230,000 new homes were built in the countryside. Between 1975 and 1980, thirty-five thousand owner-occupied houses have been constructed. More than 70 percent of the homes of employees in agriculture are privately owned; 75 percent of them are one-family or two-family dwellings.

In order to illustrate the tendencies in the period after 1971, the situation in the community of Leimbach, in the district of Querfurt, can be mentioned. The district of Querfurt has fertile soil, a comparatively low level of industrialization, and poor housing conditions. There are 470 inhabitants in the community of Leimbach, and 31 percent of the working population is employed in agriculture. The village has a co-op shop, a bakery, two pubs, a club, a nursery and kindergarten, and a medical center with a nurse. Schoolchildren are taken by bus to the neighboring village of Lodersleben. In 1971, the 138 homes had the following conditions (classified from 1 to 4 according to decreasing quality): 1 (11 homes), 2 (116 homes), 3

(11 homes), 4 (none). Of the 138 homes, 47.8 percent were built before 1870, 26.1 percent between 1870 and 1918, 3.6 percent between 1919 and 1945, and 22.5 percent between 1946 and 1970. Between 1971 and 1976, seven houses, with eleven apartments, were built. Various facilities were improved in the same period: central heating (from 2 homes to 8), water mains (from 85 to 128), inside toilet facilities (from 25 to 74), bathrooms/showers (from 43 to 75), and sewage disposals on the premises (from 47 to 91). Plans for 1980 include a central water supply system; the renovation of two apartments inhabited by old-age pensioners; the equipping of five flats with central heating; the building of a multipurpose house with bowling alley and youth club; and the improvement of the road conditions in the community. Every household in one way or another contributes twenty-five man-hours to these plans. (2) The community of Leimbach exemplifies the successful efforts made to improve living conditions in the countryside.

If one looks at regional development since the last census, the developments in the counties of Neubrandenburg and Rostock are particularly striking. The "agrarian" county of Neubrandenburg is an excellent example of the successful efforts of the socialist state in the removal of existing differences between town and country. Positive change has also been achieved in rural areas near large cities and industrial districts. The urban communities contribute to the improvement of working and living conditions not only with their higher degree of infrastructural provision, but also with the extension of central supply and other networks. This can be put into effect much more economically than in comparatively thinly populated areas far away from big cities.

Also, the cultural and social progress in the countryside is remarkable. In 1945 there were still 4,114 one-class schools on the territory of the present GDR (more than half of them in villages); the last was closed at the end of the 1959/1960 school year. The peasants' children as well as children of the working population received training facilities and educational opportunities unknown before the 1950s. The proportion of full-time employees with complete vocational or professional training was only 5.9 percent in 1960, as compared with 85.4 percent in 1978. Highly specialized production is unthinkable without an equally high level of education. Prominence must be given to the enormous progress made in the training of female staff. By 1978, 82.1 percent of women were fully qualified in a trade or profession.

With the transition to co-op production, for the first time in their lives, the peasants had the opportunity to go on a holiday. The daily working hours per full-time employee were cut from 11.6 hours in 1954/1955 to 6.7 hours in 1976/1977 (over a seven-day work week). Many sociopolitical measures made the work of female co-op members easier (among other things, there are 892 out of 1,000 children of preschool age in well-equipped kindergartens or nurseries).

The income of those working in agriculture has reached the level of industrial workers. Social security in cases of illness, accident, or old-age is guaranteed to all co-op farmers. Up to 1977, 404 outpatient clinics were established, and there are now 5,121 medical points with nurses.

Under the leadership of the Socialist Unity party, the working and living conditions of the rural population were considerably improved. This successful revolutionary change in living conditions is one of the greatest achievements in the thirty-year history of the socialist German state.

NOTES

1. Program of the Socialist Unity Party of Germany. (Berlin: Dietz, 1976), p. 30.

2. S. Haupt. "Analyse der Wohnbedingungen in der AIV Querfurt auf der Grundlage der Volks-, Berufs-, Wohnraum- und Gebaudezahlung, 1971" (Analysis of Living Conditions in the region of Querfort on the basis of Population, Professions, Living Space and Buildings, 1971), Diploma Papers, Crop-Production Section, (Martin Luther University, Halle-Wittenberg: 1978).

Anders, H. and W. Roubitschek, "Standortverteilung der Landwirtschaft und Nahrungsguterwirtschaft" (Territorial Distribution of Agriculture and Foodstuff Production), in Okonomie der sozialistischen Landwirtschaft und Nahrungsguterwirtschaft" (Berlin: 1980).

Kummer, L. and W. Roubitschek, "Die Differenzierung der Siedlungsstruktur in den Agrargebieten der DDR: eine Auswetung der Volks-, Berufs-, Wohnraum- und Gebaudezahlung, 1971" (Differentiation in the Settlement Structure of the Agricultural Regions of the GDR: An Evaluation of Figures of Population, Professions, Living Space and Buildings), Wiss. Beitrage der MLU, 12, 1977: 41-53.

Meuer, A. and W. Roubitschek, "30 Jahre DDR – erfolgreiche Entwicklung der sozialistischen Landwirtschaft" (30 Years of the GDR – The Successful Development of Socialist Agriculture), Zeitschrift fur den Erdkundeunterricht, 8-9, 1979: 305-318.

Roubitschek, W., "Abschnitt Land- und Nahrungsguterwirtschaft" (Agriculture and Foodstuff Production), in Oekonomische Geographie der DDR (Leipzig/Gotha: 1977).

Roubitschek, W., "Standortprobleme der Getreideproduktion" (Territorial Problems in Cereal Production), Wiss. Beitrage der MLU, 6, 1979: 142-156.

Sachse, D., et. al., Zur Agrar- und Bundnispolitik der SED bei der Gestaltung der entwickelten sozialistischen Gesellschaft (On the Agricultural and Alliance Policies of the Socialist Unity Party of Germany in a Developed Socialist Society) (Berlin: 1978).

4 Agriculture and Rural Transformation in Puerto Rico: Problems and Prospects

Barry Floyd

The Commonwealth of Puerto Rico, a Caribbean state with unique political and economic ties with the United States, has long been viewed as a showcase of development for other emerging countries in the West Indies, Latin America, and elsewhere. Living conditions for the majority of the island's inhabitants, particularly over the past thirty years, have clearly been enhanced as a result of the American connection. Puerto Rico has currently the highest GNP per capita in all of Latin America. Yet, in 1980 the country is facing complex social and economic problems which threaten the well-being of the state.

These problems stem in part from her dependency relationship with a powerful neighbor, but they are due also to the misconception and misdirection of development policies originating within the country. Since these policies are now being emulated by other Third World states, the Puerto Rican experience should serve as a warning to those countries seeking development, of the difficulties they too may face unless alternative strategies are adopted.

With hindsight, the relative neglect of the agricultural sector in favor of the "uphill push" toward industrialization appears largely responsible for the present social and psychological malaise on the island.

Apart from a short phase of gold mining in the early years of European colonization, agriculture was the dominant activity for most Puerto Ricans until around 1955. During the first four centuries

of Spanish settlement, the small population created
an adequate rural-based economy in which subsistence farming satisfied local needs while the specialized crops of ginger, sugar cane, coffee, and
tobacco were produced for export to the mother
country. While the island is small (only 179
kilometers by 60 kilometers) and mountainous, with
a wet/dry tropical climate and ecosystem, the
fertile coastal plains and interior basins - with
an unusual range of over four hundred soil types
and phases - encouraged a rich mix of crop and
livestock farming patterns (see figure 4.1). Many
of Puerto Rico's agricultural products derive from
plants and animals introduced from the Old World:
for example, coconuts, bananas, plantains, yams,
sugar cane, coffee, ginger, and cattle. Indigenous
plants include maize, cassava, sweet potatoes,
tanniers, pineapples, and other fruits.

The cultivation of cane and coffee provided
the main source of export revenues during the late
Spanish period. By the end of the nineteenth
century, the volume of sugar and by-products (rum,
molasses) were valued at $4.7 million. Over fifty
million pounds of coffee were harvested in 1897,
valued at $7 million. In the same year, production
of cigar and chewing leaf tobacco, largely for
domestic consumption, was valued at $600,000. The
agricultural economy of Puerto Rico on the eve of
American occupation was basically sound. While not
producing a great abundance of foodstuffs, there
was no overwhelming poverty or malnutrition among
the island's small farmers, and the country was
self-sufficient.

Puerto Rico's economy changed dramatically
after the United States assumed control in 1898.
Large corporations proceeded to acquire the best
agricultural land in the coastal plains; soon the
country was experiencing the same sugar latifundia,
or plantation economy, that was afflicting most
other countries in the Caribbean. A United States
congressional law in 1900 (reenacted in the Second
Organic Act of Puerto Rico, 1917) forbade a corporation from owning over five hundred acres of land,
but this law was blatantly ignored or circumvented.
Sugar enjoyed protection under a United States
tariff, so that cane spread over the valleys and up
the hillsides like wildfire. The small landholders
of the Spanish period were swallowed up by estates
several thousand cuerdas in extent (1 cuerda = 0.97

acres or 0.39 hectares). Multimillion dollar mills
were erected, and tight American control of the
surrounding countryside ensued. In the absence of
protective tariffs on the United States markets,
there was a drastic decline in the production of
coffee and tobacco.   The acquisition of land for
cane also induced a drop in the production of crops
used for local trade and domestic consumption.

During the first half of the present century,
a rural economy of adequacy was thus transformed to
one of deprivation for the mass of the peasantry.
At the same time, improved health measures, humane-
ly introduced by the American authorities, led to a
doubling of population within thirty years.   The
new emphasis on monoculture and on the acquisition
of lands formerly devoted to subsistence cropping
initiated the introduction of food imports from the
mainland,   which   over   the   years   have   risen   to
disturbing and costly levels.

During the two world wars, with the threat of
blockade from German submarines, under government
incentives,   the   island   produced   more   for   home
consumption.   But the postwar pattern soon returned
to one of plantation export agriculture and imported continental foodstuffs.   In 1941 there were
renewed efforts to eradicate the corporate latifundia. By that year, fifty-one sugar companies had
come to control 249,000 acres of the best farmland
in Puerto Rico.   One company alone held more than
fifty thousand acres.   The fresh attempt at land
reform was only partially successful.   Of thirty-
three companies owning more than five hundred
acres,   only   seven   were   made   into   public   enterprises.   Five others were persuaded to sell part or
all of their land to colonos; these were farmers
growing cane on the periphery of the estates, cane
which   was   then   processed   in   the   company   sugar
mills.   Further efforts to implement the Five Hundred Acre Land Law were abandoned by the end of the
decade.

Also in 1941, attention turned to the plight
of small farmers and landless agricultural workers.
The Land Act was passed, which, under a number of
titles,   proposed   various   resettlement   schemes.
Sugar plantations acquired by the government were
converted   to   so-called   proportional   profit   farms
(Title IV), whereby paid laborers were recruited to
work 100- to 500- acre cane farms under government
managers, with one-acre holdings for personal food
production.   There   were   no   residential   or   social

arrangements, and the scheme lost millions of dollars in the ensuing years due to poor management and a disinterested work force. Title V of the Land Act proposed the creation of small commumities of resettled agregados (former cane cutters), cultivating "parcels" of land of between 250 and 1,000 square meters as private family plots. Some sixty-five thousand families were to be involved, grouped in 363 planned cooperative settlements for the purposes of supervision. Three supporting agencies were given responsibility for the parcelas program: the Land Authority and Social Program Administration, the Puerto Rican Development Administration, and the Department of Public Works. This well-intended scheme floundered in the 1950s for administrative and political reasons. Today the settlements are simply small, low-income, urbanized communities located throughout the rural areas of Puerto Rico.

Under Title VI of the 1941 Land Act, the establishment of some 1,477 viable, small family farms (minifundia) of between 15 and 35 acres was proposed. Various kinds of production and social assistance from PRDA and the Soil Conservation Service were to be offered to the settlers, who were screened for motivation and experience. Initially, this program met with somewhat greater success than those under other sections of the act. But as time passed, the allocation of family holdings became a means of granting political awards, while expropriation could be inflicted as a punishment. Title VI projects are still in force, though operating under modified criteria and under the wing of the Rural Development Corporation.

Despite the attention directed toward improvements in agriculture, far greater emphasis was placed after 1948 on industrial development, under the well-publicized program known as "operation bootstrap." With massive government funding, an ambitious attempt was launched to attract United States manufacturing firms to Puerto Rico. The Economic Development Administration (EDA, or Formento) was created in 1950 and charged with industrial promotion. A wide range of schemes was adopted, including generous tax-free "holidays" for new companies, low-cost loans, prefabricated factories, utilities, cheap labor, and personnel and management training programs. These incentives virtually eliminated many of the risks of initial

investment and enhanced the possibilities of profitable operations.

The successes of Formento in the 1950s and
1960s were impressive. Large-scale industrialization proceeded to transform the Puerto Rican economy. Hundreds of factories were erected, the basic
infrastructure was provided, and job opportunities
were created for some eighty-seven thousand workers
in the first decade of the program. Nondurable
consumer goods - were among the earliest products.
In 1955, the income from manufacturing exceeded
that from agriculture for the first time. The
widespread internal migration from rural to urban
areas of persons in search of industrial employment
led to labor shortages on the land and a further
decline in agricultural output. In figure 4.2
official statistics of agricultural production in
Puerto Rico in 1965/1966 and 1971/1972 reveal the
deteriorating situation. Figure 4.3 gives an indication of the mounting dependence on imported
foodstuffs between 1950/1951 and 1970/1971.

In the 1960s Formento began attracting heavier
industries - chemicals, electronics, and metallur-
gical products - in efforts to diversify and to
create a broader, securer base for the manufacturing sector. A petrochemical complex provided over
sixty thousand jobs on full stream; unfortunately,
it was built on land with high agricultural potential on the south coast. In the 1970s, international pharmaceutical companies began locating in
Puerto Rico in order to penetrate the continental
United States market.

Nevertheless, in the face of the energy crisis, the economic recession in the United States,
and fragile financial situations elsewhere, the
vulnerability of Puerto Rico to world business
cycles has been exposed. Recently, a number of
American companies have closed down their operations on the island, in response to mounting labor
costs related to minimum hourly wage legislation on
the mainland - valid also in Puerto Rico.

In spite of undeniable economic progress over
the past thirty years, the country had a large
trade deficit in 1977. Imports of agricultural
commodities from the United States alone amounted
to over six hundred million dollars. An estimated
70 to 80 percent of food being consumed in Puerto
Rico was originating from overseas producers; a
dangerous economic dependence, particularly in view

of the increased incidence of dock strikes and disrupted shipping movements, reminiscent of World War II.

The contribution of the agricultural sector to the gross national product was a mere 10 percent, while the labor force in farming was down to a bare 5 percent. At the same time, unemployment in Puerto Rico was officially 22 percent (unofficially, over 30 percent), out of an adult labor force of 960,000 and a total population of 3.28 million. Seventy percent of the island's families have annual incomes below the minimum deemed necessary by the Commonwealth government to maintain an adequate standard of living in Puerto Rico. Thus, almost three-quarters of the population are in receipt of food stamps, a social security device which enables beneficiaries to buy essential (imported) foodstuffs at much reduced prices. In the United States only 5 percent of the population receives food stamps.

In contemporary Puerto Rico there are serious inequities in the distribution of wealth; 45 percent of the population receives only about 15 percent of generated incomes. The crime rate is mounting and drug addiction is widespread. Substandard housing in both town and country is much in evidence (see figure 4.4).

Crass materialism and selfish behavior born of the consumer society are eroding the traditional values of the family and the Spanish-American cultural institutions on the island. The frustrations of men and women from many walks of life are a dominant characteristic of present Puerto Rican society. In consequence, political dissention at the grass-roots and national level is rife.

One solution to the current socioeconomic crisis in Puerto Rico is to revitalize agriculture and the rural communities, avoiding at the same time the bureaucratic errors of earlier measures to aid small farmer societies. This will not be easy in the Puerto Rican context. In the 1970s there has been a proliferation of agricultural planning groups, projects, and agencies in efforts to aid the sick man of the economy. The ailing Alexander the Great is said to have protested that he was dying with the aid of too many physicians. If the agricultural sector of Puerto Rico could talk, it might say as much. Nevertheless, there are encouraging signs that radical rethinking about the

productive and stabilizing role of an enlightened
small-holder work force, harvesting products for
domestic consumption as well as for export under
modern methods, and with incomes comparable to
those working in urban-based factories, will provide a valuable sheet anchor for the future.

As indicated, the sad state of Puerto Rican
agriculture in the 1970s was not due to the complete neglect of the sector in favor of all-out
industrialization, with the ultimate assumption
that all food for the island would be imported from
abroad. But injections of American and foreign
capital into manufacturing enterprises, together
with neglect of zoning regulations, unchecked urban
sprawl, and land speculation (there are over seven
thousand real estate development corporations on
the island) have, over the years, dealt crippling
blows to indigenous rural land use. The sad decline in farming is a symptom of the malady caused
by overdependence. In truth, agriculture has over-
long been the ugly duckling of the Puerto Rican
economy.

Concerned with the state of agriculture, the
Governor's Advisory Council for the Development of
Government Programs turned its attention to the
sector in 1970. Recommendations for tighter land
classification were advanced, with the preservation
of good-quality agricultural land as a high priority. Improvements in public works, such as feeder
roads, bridges, drainage, and irrigation, were
sought. Increased incentives for production, such
as improved housing, subsidies, agricultural credit
and insurance, better seeds, cheap fertilizer,
appropriate technology, and demonstration farms,
were urged. A new marketing procedure for traditional foods - farinaceous crops, fruits, and
vegetables - was strongly endorsed, with encouragement of cooperatives for more efficient transportation, storage, processing, and marketing. These
and other proposals to rationalize domestic food
production, while technically and economically
sound, nevertheless require full-scale govenment
backing in terms of investment and trained manpower. In reality, these have not been forthcoming
at anywhere near the levels required, while excessive centralized bureaucracy has hindered guidance
and funds from trickling down to the very beneficiaries for whom the improvements are planned: the
small farmers.

In 1972 the Junta de Planificacion, Oficina
del Gobernador, Estado Libre Asociado de Puerto
Rico produced a master plan for agricultural land
use in the country. Existing conditions and tendencies were reviewed, options for the future were
analyzed, and specific areas for agricultural development were identified. The seriousness of the
situation was emphasized:

> The crisis of the agrarian sector has
> prompt and fatal repercussions on the activ-
> ities of life. The best guarantee of a strong
> economic potential is the existence of a
> progressive and widespread agriculture which,
> if possible, reaches self-sufficiency. It is
> absolutely necessary to cultivate the land so
> that the highest returns are created from
> agrarian products....
> The widespread effect of agrarian activity
> is highly important. There are two reasons
> why these multiple effects are greater in
> agriculture than in other sectors. Firstly,
> by nature, agrarian activity does not require
> substantial imports for its production. Agri-
> culture is fundamentally a consumer of goods
> and services that are produced locally. This
> type of consumption reduces the leakage of
> income out of Puerto Rico. This does not
> happen in the manufacturing sector which re-
> quires a high percentage of imported raw
> materials in order to operate.
> In second place, the profits that are
> generated by agrarian activity nearly always
> remain in Puerto Rico. The agriculturalists
> live on the island and spend the money there.
> (1)

In 1973 the Corporation for the Rural Development
of Puerto Rico (CDR) was created, together with the
Administration for the Development of Agricultural
Enterprises (AFDA). The statement of objectives
for the CDR acknowledges with unaccustomed frank-
ness the shortcomings of previous efforts to aid
agriculture:

> The programme for economic improvement
> tried out for the rural areas by the Govern-
> ment of the Free Associated State have not
> produced all the benefits that were expected

because they set out with the assumption that
agrarian activity could be promptly rehabili-
tated and converted into an economically productive activity. That is not always possible. The lack of units of production of an
adequate size; the low yields of the crops
that are grown; the difficulty in mechanizing
cultivation and harvesting; the scarcity of
capital and financial aids; the lack of human
resources adequate for agrarian labor; the
deficiency of the marketing systems linked to
the poorly developed means of transportation
and communication. All these factors combine
with the concomitants of the cultural features
to produce a discouraging perspective of the
rural area and people of Puerto Rico. (2)

The Rural Development Corporation has responsibility for the Title VI program in reactivated
form. Capital-intensive poultry-breeding units are
being promoted on 10- to 15-acre farms, with eggs
flown in from Florida, incubated in Puerto Rico,
and the chicks reared on imported feedstuffs.
Vegetables are also being grown as cash crops for
the local market. But high investment costs and
long-term loans benefit only a few privileged
farmers. The selection of candidates for Title VI
farms and conditions of usufruct continue to create
difficulties. CDR has also revived the work of the
Commission for the Improvement of Isolated Communities, established in 1960 to try to aid some of the
poorest and most disadvantaged of rural dwellers in
mountainous areas. There is thus a strong social
welfare component to the work of the Rural Development Corporation, with its focus on small farmers.
    AFDA, on the other hand, is charged with
developing the economic infrastructure of rural
areas, aiding middle-sized farms, promoting more
efficient marketing, and encouraging processing
plants and agro-industries. The need to promote
import substitution of foodstuffs, reversing the
flow of expensive foreign foods (or, in some cases,
dumped surpluses of American products) is clearly
recognized, in a world where competition for food
beyond one's own borders will be all too common in
future.
    The separation of functions between CDR and
AFDA is not without its limitations, however.
According to one critic, the formal demarcation of
two sister corporations, one with a 'social' focus

and the other with an 'economic' focus perpetuated
an already mythological differentiation between a
marginal family-based agriculture and an 'efficient' commercial agriculture. (3)

Meanwhile, additional allocations of small
holdings to needy and deserving farmers, and even
to urban dwellers anxious to break away from the
strictures of industrial employment (and unemployment), have proceeded sluggishly. The Program of
Land Distribution (PDT) is dismissed as the deni-
grated Cinderella of the agro-rural policy (4) by
one rural sociologist with a wide knowledge of land
reforms in Puerto Rico.

In late 1975 the Development Strategy Committee of the Government Development Board produced a
remarkable document entitled A Plan for the Parallel Economy of Self-Help Cooperatives in Puerto
Rico, soon referred to as the Vietorisz and Goldsmith report, after its principal authors. The
study acknowledged that even if conventional strategies for boosting the economy in both industrial
and rural sectors were successful, the magnitude of
foreseeable unemployment (between 100,000 and
300,000) would still be unmanageable. With the
shaky American economy, large-scale migration of
Puerto Ricans to the continental United States can
no longer be counted on to relieve population
pressure, as in earlier decades. Indeed, a reverse
flow has been a characteristic of recent years.

The Vietorisz and Goldsmith report proposed
the creation of a so-called integrated parallel
economy - a set of cooperative and largely self-re-
liant farms with complementary factory and service
enterprises that would use minimum government investment, social programs, or other aid, but which
would produce most of their own needs (such as
housing and food) and whose members would become
trained and educated for efficient and rewarding
participation in the "normal" economy.

The fundamental supposition behind this idea
is that nothing is so wasteful and inefficient
as forcing a person to remain idle while his
own and his family's needs go unprovided for.
If no jobs are available, this person must be
helped to help himself. Traditional governmental aid programs - welfare and others - do
provide somewhat for needs, but they ignore
the problem of idleness, and they ignore the
need for production. Yet a surprising number

of active and productive things can be done, as attested by the widespread do-it-yourself movement of home and appliance repairs, the making of household formulations and preserved foods, even the building of houses. The idea is to generalize this group of self-help activities and apply the concept to a range of activities that jointly can make a substantial difference to the real level of living of a poor family. That which is not possible individually can be done by people jointly. That which is technically inefficient, and many of these activities will be highly inefficient by formal economic calculus, at least in the initial stages, is nevertheless highly useful: some production with totally wasted resources is better than none, and pleasurable activity is better than unhappy idleness or alienation. (5)

In an appendix to the Vietorisz and Goldsmith plan, there are detailed notes on community planning for self-help cooperation, complete with elaborate graphic illustrations of feasible layouts and suggested spatial arrangements for residential areas, learn-work and work-learn activities, commercial, cultural, and recreational zones.

In sum, here is official endorsement for a policy of self-help cooperatives or communes that are an increasingly common "alternative life-style" feature of rural areas in the industrialized societies of North America and Western Europe. In 1976 the government declared that the "parallel economy" would be its principal weapon for combatting unemployment. But, in a semantic switch, the "cooperatives of self-aid" were transposed into "training and labor camps" (a sort of domestic Peace Corps) and placed under the direction of the Department of Public Instruction. The possibility of using the young unemployed as a vehicle for radical rural development seems to have receded. Only in a few localities, the most noteworthy being the Union de Pequenos Agricultores de Carite, have attempts been made by private organizations (though aided by federal funds) to create viable models of self-help communes in Puerto Rico. An enterprising feature of Carite's program is the growing of ornamental plants and shrubs for sale to gardening enthusiasts at home and abroad.

It is still too early to speak of a "new era

in agriculture", a phrase favored by the veteran geographer of Puerto Rico, Rafael Pico. A new, far-reaching and properly enforced agrarian reform law is needed to achieve a more equitable distribution of land for full-time professional farmers. There must be an end to unchecked encroachment of the urban environment on agriculturally productive land. Appropriate technology in terms of miniaturized equipment and improved package inputs for small farmers will encourage more scientific and productive tillage and animal husbandry. Bilateral agreement with the United States government to protect Puerto Rican agriculture from mainland vertically integrated agro-industries is vital.

In the last analysis, a deliberate change in the direction of national economic planning is called for. In the words of a young Puerto Rican geographer, Angel David Cruz Baez:

It has been proved...that industrialization per se, especially when it is based on the importation of capital, raw materials, technology and also directed towards international markets is not the panacea that will resolve all the economic problems of the Third World. Industrialization has already been given its opportunity and we can see where it is leading Puerto Rico. It will not be able to go on for much longer without causing more environmental, health and economic problems than it has caused already. Why do we not now give an opportunity to a scientific and efficient agriculture that will act in response to the interests of Puerto Rico as a nation? (6)

NOTES

1. Estado Libre Asociado de Puerto Rico, Oficina del Gobernador, Junta de Planificacion, Area de Planificacion Fisica, Puerto Rico: Plano Regulador Uso De Terrenos, Sector Agricola (San Juan, 1972): 3-5. (English translation)

2. Estado Libre Asociado de Puerto Rico, Ley 63 del 30 de mayo de 1973 (P. del S. 156. Conferencia): 2. (English translation)

3. Franscisco Watlington-Linares, "El Reto del Desempleo y la Politica de Distribucion de Tierras en Puerto Rico," Revista de Ciencias Sociales, 14, 4 (Diciembre, 1975): 373-4. (English translation)

4. Ibid., p. 374.

5. Thomas Vietorisz and William W. Goldsmith, A Plan for the Parallel Economy of Self-Help Co-operatives in Puerto Rico, paper prepared for the Development Strategy Committee of the Finance Council, Government Development Bank for Puerto Rico (Nov. 10, 1975): 1-2.

6. Angel David Cruz Baez, "'Nueva era agricola' o Sindrome de Deterioro Agricola: El Comportamiento de la Agricultura en Puerto Rico en los Ultimos Anos," Revista de Ciencias Sociales, 18, 3-4: 38. (English translation)

REFERENCES

Government Publications

Commonwealth of Puerto Rico, Office of the Governor, Puerto Rican Planning Board, Urban Land Policy for the Commonwealth of Puerto Rico. San Juan, 1968).

Governor's Advisory Council for the Development of Government Programs, Agriculture. (San Juan, 1970).

Commonwealth of Puerto Rico, Rural Development Corporation of Puerto Rico, Department of Agriculture, Regulation to Reenact the One Approved on October 31, 1975, Entitled "Regulation for the Distribution and Operation of Farms Granted under Title VI of the Land Law." (Santurce, Puerto Rico, 1976).

Estado Libre Asociado de Puerto Rico, Oficina del Gobernador, Junta de Planificacion, Objetivos y Politicas Publicas del Plan de Usos de Terrenos. (San Juan, 1977).

Estado Libre Asociado de Puerto Rico, Departamento de Agricultura, Oficina de Estadisticas Agricolas, Ingreso Agricola de Puerto Rico, 1975/76 - 1976/77. (Santurce, Puerto Rico, Oct. 1977).

Other Publications

Badia, R.M., "Puerto Rico," in World Atlas of Agriculture, (Novara, Italy: 1969): 388-95.

Balchin, W. and Coleman, A., "Puerto Rico," in Geography, 50, 228 (July 1965): 274-86.

Bonilla, Eduardo Seda, Social Change and Personality in a Puerto Rican Agrarian Reform Community (Evanston, Illinois: Northwestern University, 1973).

Bourne, D.D. and R. Bourne, Thirty Years of Change in Puerto Rico: A Case Study of Ten Selected Rural Areas (New York: Praeger, 1967).

Carron, Alain-Marie, "Puerto Rico: Island in Search of an Identity," in The Guardian (October 23, 1977): 13.

Duggal, Ved P., "Decline of Agriculture in Puerto Rico: Agriculture in a Small Economy Industrializing with the Help of Foreign Captial," Proceedings of the Ninth West Indian Agricultural Economics Conference (Trinidad: University of the West Indies, 1974): 32-42.

Howard University, Symposium: Puerto Rico in the Year 2000 (Washington: Howard University Press, 1968).

James, Mike, "Puerto Rico - No Model for Us," Caribbean Contact, 5, 12 (April, 1978): 17.

Jones, Clarence F., (ed.), The Rural Land Classification Program of Puerto Rico (Evanston, Illinois: Northwestern University Studies in Geography, No. 1, 1952).

Land Tenure Center, Agrarian Reform in Latin America: An Annotated Bibliography, Land Economics Monographs No. 5 (Madison, Wisconsin: University of Wisconsin, 1974).

Lewis, Gordon K., Puerto Rico: Freedom and Power in the Caribbean (New York: Monthly Review Press, 1963).

Lewis, Oscar, La Vida: A Puerto Rican Family in the Culture of Poverty, San Juan and New York (New York: Random House, 1966).

MacPhail, D., "Puerto Rican Dairying: A Revolution in Tropical Agriculture," Geographical Review, 53, 2 (1963): 224-46.

Mathews, Thomas, "The Agrarian Reform in Cuba and Puerto Rico," Revista de Ciencias Sociales, 4, 1 (March, 1960): 107-23.

Mayne, Alvin, Designing and Administering a Regional Economic Development Plan, with specific reference to Puerto Rico (Paris: OECD, 1961).

Mendez, Eugenio Fernandex, (ed.), Portrait of a Society: Readings on Puerto Rican Sociology (Rio Piedras: University of Puerto Rico, 1972).

Mintz, Sydney W., Worker in the Cane: A Puerto Rican Life History, Caribbean Studies No. 2 (San Juan: Institute of Caribbean Studies, 1969).

Niddrie, David L., "The Problems of Population Growth in Puerto Rico," Journal of Tropical Geography, 20 (1965): 26-33.

Pico, Rafael, The Geography of Puerto Rico, (Chicago: Aldine, 1974).

Rivera, L. Santos, "Tenure Innovations and Agricultural Production in Puerto Rico," in K.H. Parsons, R.J. Penn, & P.M. Raup, eds., Land Tenure (Madison, Wisconsin: University of Wisconsin Press, 1956): 328-37.

Roces, M. Villar, Puerto Rico y su Reforma Agraria (Rio Piedras: Editorial Edil, 1968).

Steward, Julian H. et al., The People of Puerto Rico: A Study in Social Anthropology (Urbana

Illinois: University   of   Illinois   Press,
1956).

Tata, Robert J. and David R. Lee, "Puerto Rico:
Dilemmas of Growth," Focus, 28, 2 (Nov./Dec.
1977).

Tugwell, Rexford, The Stricken Land: The Story of
Puerto Rico (New York: Doubleday, 1947).

United States Bureau of the Census, Census of the
United States: Agriculture of Puerto Rico
(Washington: GPO, 1970).

Universidad de Puerto Rico, Colegio de Ciencias
Agricolas, Estacion Experimental Agricola, Es-
tudio   Socio-Economico del Programa   de Fincas Individuales del Titulo VI de la Ley de
Tierras, Boletin 236 (Rio Piedras: April,
1974).

Universidad de Puerto Rico, Colegio de Ciencias
Agricolas, Estacion Experimental Agricola, Es-
tudio   Socio-Economico de 29 Fincas Individuales del Titulo VI de la Ley de Tierras Dis-
tribuidas de 1967 a 1970, Boletin 241 (Rio
Piedras, Puerto Rico,: Febrero, 1976).

Wagenheim, Kal, Puerto Rico: A Profile (New York:
Pall Mall Press, 1970).

Watlington-Linares, Francisco, "Nuevas Tendencias
en la Dinamica Social de la Communidad Agro-
rural," Revista de Ciencias Sociales, 14, 1
(March, 1975): 24-45.

Wells,   Henry,   The Modernization of Puerto Rico
(Cambridge, Massachusetts: Harvard University
Press, 1969).

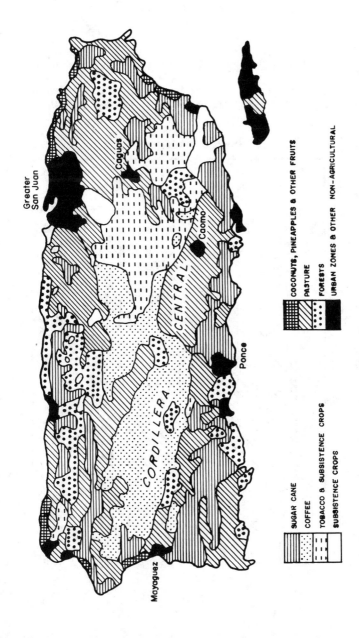

Figure 4.1. Puerto Rico: Land Use.

Source: After H. Berrios, Puerto Rico Planning Board.

61

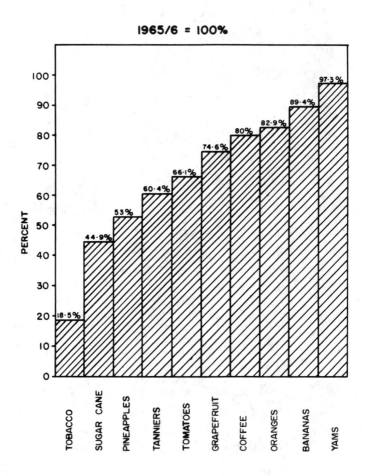

Figure 4.2.   Decline in Output of Specific Crops in
Puerto Rico: 1965-1966 to 1971-1972.

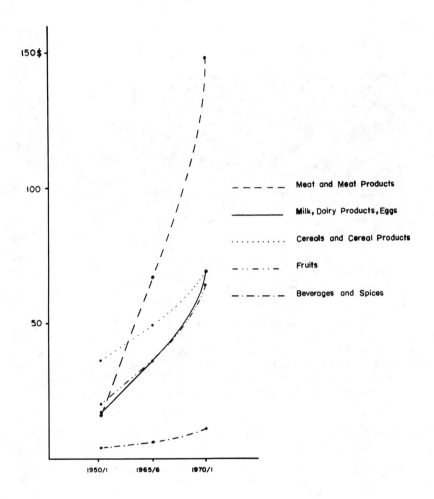

Figure 4.3. Comparative Values of Food Imports into Puerto Rico (Nearest Million US dollars in Current Prices).

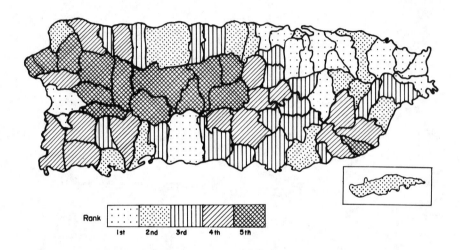

Figure 4.4.   Puerto Rico: Quality of Life.

Quality of life scores are based on median family income, median school years completed, percentage of employed males not engaged in farming, population change 1960-1970, and condition of housing.

Source: After R. Tata and D. Lee.

5 Agricultural Change and Rural Stability in Australia

Peter Scott

Australia is characterized by rural stability rather than rural transformation. Although the agricultural sector is undergoing continual change, Australia has never had a peasant society, and throughout its history the dominant unit of agricultural occupance has been the owner-occupied family farm. Its agriculture, which has evolved for the most part in a dry environment with marked climatic variability and exceptionally poor soils, is highly specialized with large-scale pastoralism and mechanized grain farming predominant. Pastoral properties include some of the largest in the world, and the highly mechanized farms are significantly more capital intensive than its factories. Even the comparatively small-scale farming of the humid southeastern and southwestern coastlands and of the irrigated settlements in the Murray-Darling Basin is essentially capital intensive and, despite comparatively low productivity per farm and per worker, furnishes high returns per unit of capital input. Yet the continued presence in these regions and, to a lesser extent, in some drier areas of low-income farmers has prompted some recent structural adjustment in Australian agriculture, with consequent localized repercussions on rural settlement patterns and rural social structure. Nevertheless, the relatively large size of most Australian farms when first established and the relatively elastic demand for many Australian farm products has led to less pressure, at least until recently, for agricultural adjustment than in Europe and North America. (1)

65

Rural stability stems essentially from the maturity and commercialization of farming, which permit substantial agricultural changes to be accommodated without concomitant changes, at least in the short term, in either the size of the rural population or the settlement pattern. Since most agricultural systems are export oriented and subject to the vagaries of overseas markets, they have developed highly flexible structures. While agriculture continues to display a substitution of capital for labor, an ever-changing mix of farm enterprises in response to changing cost-price relationships, and an increasing scale of activity, major structural change is confined to few regions. These regions are mostly those that have specialized in horticulture or dairying and have been adversely affected by Britain's entry into the European Economic Community. Otherwise, the reduction in farm numbers tends to be most prevalent near metropolitan centers as a result of urban encroachment. On the other hand, farm numbers have increased in many regions where either land formerly considered unsuitable for agriculture has been developed in response to scientific and technological advance or agricultural land has been subject to closer settlement in response to changing market conditions.

Australia's rural population may therefore be said to display stability rather than exodus. (2) Although the movement of mostly young people from the country to the towns and particularly from small rural service centers to the larger cities continues to be pronounced, the interaction of net outward migration and reduced rates of natural increase tends to give a relatively stable population in most rural areas. Moreover, net migration involves not only substantial outflows but also significant inflows: in the 1966-1971 period, for instance, rural-urban flows accounted for only 8 percent of all internal migration while urban-rural movements accounted for 6 percent. At the same time, rural-rural movements made up 5 percent and urban-urban movements, indicative of a high level of urbanization, no less than 81 percent. Furthermore, internal migration embraced a higher proportion of the urban than of the rural population. In short, rural population movements would seem to be a largely equilibrating mechanism for maintaining rather than changing the size of local or regional populations. (3)

This chapter focuses on recent changes in farm numbers and the rural population against a background of the changing enterprise structure of Australian farming in response to changing costprice relationships. It pays particular attention to the intercensal period 1966–1971 when the rural sector underwent the greatest upheaval of any postwar intercensal period. In these years, when the market orientation of Australia's agricultural exports shifted from Western Europe to East and Southeast Asia, farm incomes experienced an acute cost-price squeeze that prompted or accelerated agricultural change, the number of commercial farms fell sharply, and the rural population recorded a substantial loss. Changes in the enterprise structure of farms and in farm numbers can be analyzed using comparable data only for five years within the period 1959–1960 to 1973–1974 and detailed data on internal migration exists only since 1965–1966. Accordingly, the years 1966 to 1971, a critical period in the postwar history of Australian agriculture, constitute the only intercensal period for which all three changes can be examined in detail. This chapter concludes with a tentative spatial assessment of the degree of agricultural and population stability manifested by rural Australia in the study period.

FARMING–TYPE CHANGES

Before reviewing the changing enterprise structure of Australian farming, it is necessary to summarize briefly the basic agricultural systems within which these changes take place. In figure 5.1 the limits of rural land use are those defined by the Atlas of Australian Resources (Department of Minerals and Energy, 1973); more than one-third of the continent is too arid or too rugged for agriculture. In figure 5.2 the intensive agricultural systems comprise, firstly, horticulture and industrial crop production, located within peri-metropolitan areas, favored localities, and irrigation settlements, and secondly, intensive livestock farming, made up of dairying, beef-cattle fattening, and sheep rearing for meat and/or wool; both systems are confined to eastern, southeastern, and southwestern coastal fringes and adjoining country. Intermediate systems comprise a combination of livestock farming and cereal cultivation, chiefly for sheep and wheat, and occupy extensive belts adjoining the intensively

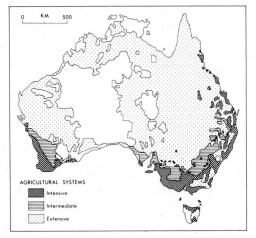

Figures 5.1 - 2.   Rural Land Use and Agricultural
                   Systems in Australia.

farmed coastal zones in the southeast and the
southwest. Extensive systems involve livestock grazing, mainly sheep in central Queensland, southern
Australia, and the northwest, and beef cattle elsewhere. Each of these systems responds to changing
cost-price relationships of its products by switch-
ing enterprises, though in some subsystems, notably
dairying and the cultivation of tree crops, change
may be inhibited by specialized capital investment
or farm size.

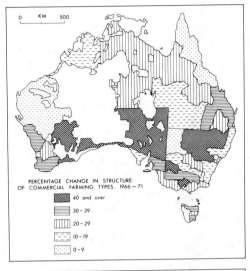

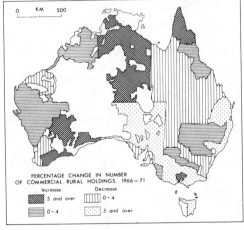

Figures 5.3 - 4.

Some insight into the changing structure of farming systems may be gleaned from official classifications of rural holdings. From 1959-1960 to 1973-1974 the Commonwealth Bureau of Census and Statistics, later renamed the Australian Bureau of Statistics, undertook five classifications of rural holdings defined as tracts of agricultural land of at least 0.4 hectares. These classifications were based on the predominant activity on each holding and were directly comparable. Gross receipts less marketing costs for each type of activity on each

holding were estimated by applying estimated values
per unit area and per livestock to areas of crops
and numbers of livestock reported on the annual
agricultural census returns. To minimize the effects of seasonal fluctuations, these estimated
values were based on averages of both yields and
prices for each state as a whole for the five years
ending in the year of classification. Perhaps the
only factor seriously limiting the comparability of
the data is the criteria used to define commercial
holdings: a minimum of estimated gross receipts
from all products of $1,200 in 1959-1960, $1,600 in
1965-1966, and $2,000 in subsequent years. But
because the percentage increase in the threshold
between 1965-1966 and 1970-1971 is less than the
rate of inflation over this period, the number of
commercial farms in 1970-1971 is, by comparison with
the earlier year, probably overstated. Accordingly,
the extent of change over the study period was probably somewhat greater than these data suggest.

     Four sets of criteria were employed to classify
holdings by type of activity. First, holdings were
designated as sheep and cereals if the combined
receipts from these activities accounted for at
least 75 percent of the total receipts of the
holding and if neither activity contributed more
than four times the other. Secondly, holdings with
cattle for milk production and pigs were classified
as milk cattle if the combined receipts from these
activities amounted to 50 percent or more of total
receipts. Thirdly, other commercial holdings were
classified as to the activity which accounted for 50
percent or more of total receipts. However, holdings were classified as "other vegetables" where 50
percent or more of total receipts were derived from
the production of vegetables of any kind, including
potatoes, but where potatoes alone did not account
for more than half the total receipts. Moreover,
holdings were classified as "other" (one main purpose) where a single product, other than those
listed in Table 5.1, such as cotton, peanuts, or
oilseeds, contributed 50 percent or more to total
gross receipts. Finally, if none of the above
criteria was satisfied, the holding was classified
as multipurpose.

     In Table 5.1 the types of activity have been
grouped under three main headings to highlight some
salient features of Australian agriculture. Holdings
specializing in cereals and oilseeds have been
allocated to crop-livestock systems because grain

Table 5.1. Number of Commercial Rural Holdings
Classified by Type of Activity, Australia
1965-1966 to 1970-1971

Type of Activity	1965-66	1970-71	% Change
Cropping Systems			
Potatoes	2,367	1,562	-33.17
Other Vegetables	6,249	6,381	2.11
Vine Fruit	4,715	4,364	- 7.44
Other Fruit	11,977	11,018	- 8.01
Sugar Cane	7,554	7,467	- 1.15
Tobacco	992	1,153	16.23
Total	33,854	31,945	- 5.64
Livestock Systems			
Poultry	4,194	3,341	-20.34
Pigs	1,576	3,084	95.69
Milk Cattle	49,334	36,486	-24.62
Meat Cattle	16,339	23,440	43.46
Sheep	43,671	47,881	9.64
Total	115,114	114,232	- 0.77
Crop-Livestock Systems			
Sheep & Cereals	31,832	32,221	1.22
Cereal Grain	12,274	7,216	-41.21
Multipurpose	8,995	3,287	-63.46
Total	53,101	42,724	-29.54
Other	1,954	1,967	0.67
Commercial Holdings	204,023	190,868	- 6.94

Source: Commonwealth Bureau of Census and Statis-
tics, Classification of Rural Holdings by Size and
Type of Activity, 1965-66, Bulletin No. 7 (Canberra,
Australia, 1968); Australian Bureau of Statistics,
Rural Land Use, Improvements, Agricultural Machin-
ery and Labour, 1972-73 (Canberra, 1974).

farming and oilseed production in Australia almost
invariably incorporate some livestock; in any case,
aside from rice-growing, they are essentially intermediate in land-use intensity by contrast with other
specified cropping systems. Unfortunately, the classification of rural holdings takes no account of the
intensity of land use; so, the numbers of holdings

specializing in either sheep or meat cattle cannot be allocated between intensive livestock breeding and extensive livestock grazing. Even so, the classification highlights, inter alia, the importance of horticulture among intensive cropping systems, the high degree of specialization among both crop and livestock farming, and above all, the preeminence of pastoralism in Australian agriculture.

Table 5.1 also illustrates some overall changes in the enterprise structure of Australian agriculture in the late 1960s. Of the three basic systems depicted, crop-livestock agriculture recorded the greatest decline in the number of classified holdings, followed by cropping; decline among livestock systems was seemingly negligible. But as livestock farming exemplifies vividly, these overall changes mask significant differential movements among specific types of activity. Thus, even though scant change occurred in the total number of livestock holdings, dairying with the contraction of European markets revealed by far the greatest absolute decline of any type of rural holding, while meat-cattle production, stimulated by American demand for beef, displayed by far the greatest absolute increase. A substantial absolute decline also took place in the numbers of multipurpose and cereal-grain holdings, while holdings engaged in sheep or pig production made considerable absolute gains. A somewhat different pattern is disclosed by the percentages of increase and decrease in the numbers of specific types of holding.

More germane to the present inquiry, however, are spatial changes in the enterprise structure of farming. Figure 5.3 shows the overall changes by statistical division, compiled from unpublished data by local government areas (many divisional boundaries were altered in the study period), kindly made available by the deputy Commonwealth statistician in the states. The greatest changes took place in the New South Wales crop-livestock belt, which, despite national trends, saw a switch from sheep to cereal grains and wheat-sheep farming, in South Australia, where sheep and wheat-sheep farming increased at the expense of cereal grains, and in the Esperance district of Western Australia, where new farms were opened up to wheat-sheep and sheep farming. Areas experiencing a lower, but above-average rate of change, such as southeastern Queensland, northeastern Victoria, and the far southwest, mostly recorded expansion in beef and sheep farming and a

decline in dairying, though the number of
cereal-grain holdings declined significantly in
parts of western Australia and Victoria. Areas with
average rates of change generally witnessed an
increase in the number of beefcattle holdings, but
this was largely at the expense of dairying in
southern Australia, where the number of sheep farms
also increased, and at the expense of sheep farming
in central Queensland. Below-average rates of change
highlight the shift from dairying to beef production
in coastal New South Wales and from sheep to cattle
grazing in southwestern Queensland. Finally, the
lowest rates of change mainly delimit the increase
in the number of beef farms along the central and
north Queensland coast, where sugar and dairy farm
numbers declined, a small decrease in sheep grazing
in western New South Wales, and a small expansion in
sheep grazing in the Carnarvon and Pilbara districts
of western Australia.

In sum, most of the change portrayed in Figure
5.3 may be attributable to the expansion of beef
production and the decline in dairying. Both these
enterprises showed consistent trends throughout the
continent, whereas sheep farming and crop-livestock
systems displayed markedly differential regional
changes. Activities not discussed above but reveal-
ing consistent increases in farm numbers are pig
raising, mainly in southeastern Queensland and northern Victoria, and tobacco growing, mainly around
Maryborough in Queensland and in northeastern Victoria. On the other hand, a consistent decline
typified multipurpose farming, notably in southeastern Queensland and southwestern Western Australia; in potato growing, chiefly also in the far
southwest; and in poultry farming, mainly around
Sydney, Adelaide, and Perth. Other activities ex-
hibiting differential movements were viniculture,
where the number of vineyards fell in Victoria but
rose in South Australia, and fruitgrowing, where the
number of holdings with temperate and subtropical
fruits declined in southern Australia while the
number of tropical fruit farms in Queensland increased.

CHANGES IN FARM NUMBERS

While changes in the enterprise structure of farming
may largely reflect the substitution of a more
profitable enterprise for a less profitable one,

changing cost-price relationships together with many
other factors, such as the reorganization of farm
businesses, agricultural innovation, and land settlement programs, may also lead to the amalgamation
or abandonment of rural holdings or to the establishment of new farms. Although the total number of
commercial rural holdings would seem to have fallen
fairly steadily since the Second World War, the lack
of data comparability precludes accurate assessment
of long-term trends; over the study period 1966-1971
the decline was nearly 7 percent (Table 5.1). In
1974-1975 the commercial rural holding, hitherto
defined as having a minimum area of 0.4 hectares,
was redefined as an agricultural establishment, with
a minimum of 1 hectare. In 1975-1976 an agricultural
establishment was redefined as having at least 10
hectares. In that year, the number of establishments predominantly engaged in agriculture was
180,400. In 1976-1977, using the same criteria, the
number was 173,650. Lette has attempted to redefine
the number of agricultural establishments from
1950-1951 to 1974-1975 using the criteria employed
by the Australian Bureau of Statistics for the two
subsequent years. (4) On this basis he gives the
number of agricultural establishments in 1965-1966
as 199,750 and in 1970-1971 as 189,400 (cf. Table
5.1); the intercensal decrease would thus be little
more than 5 percent. But whatever the precise
totals, two postwar trends are indisputable: the
total number of commercial rural holdings has fallen
considerably, perhaps by at least 15 percent since
1960; (5) and the number of part-time subcommercial
holdings, located mainly near metropolitan areas,
has increased strikingly, perhaps by as much as 50
percent since 1960.

Yet these trends relate merely to national
aggregates; over much of western and northern Australia the number of commercial rural holdings
increased in the study period, while over much of
eastern and southern Australia the number declined
(see Figure 5.4). The Western Australian wheat belt
experienced a strong demand for both old marginal
land suitable for resettlement and new pioneering
land on the northern and southern drier margins of
the crop-livestock zone. (6) But the greatest increase in farm numbers took place on the Esperance
sand plain of Western Australia, where a large-scale
development project for sheep and wheat-sheep farming attracted considerable overseas investment, particularly from the United States, and attracted

widespread among wheat-sheep farms and conducive to
scale economies, but it enables farmers to take
advantage of diverse soil conditions and to spread
environmental risks. (11) But perhaps the greatest
scale economies and even the greatest technological
advances recently achieved in Australian agriculture are those accruing to poultry farming, where
large-scale entrepreneurs vertically integrated with
feed merchants and retail outlets have entered the
industry. Consequently, the study period witnessed
a sharp decline in the number of poultry farms,
markedly so in areas around Sydney, Melbourne,
Adelaide, and Perth.

RURAL POPULATION CHANGE

A decline in farm numbers, if not the changing
enterprise structure of farming, might be expected
to contribute to the movement of labor resources out
of agriculture. Certainly, the total number of
males employed in agriculture, as recorded in the
population census, fell from 430,000 in 1954 to
384,000 in 1961, 355,000 in 1966, 298,000 in 1971,
and 257,000 in 1976. Thus, the rate of decline in
the study period was greater than in any other
postwar period. Since the total number of males
employed in Australia increased throughout these
years, the proportion engaged in agriculture, which
had been as high as 28 percent in 1933 when 492,000
were employed, fell from 15 percent in 1954 to 13
percent in 1961, 11 percent in 1966, 8 percent in
1971, and 7 percent in 1976. It follows that the
rate of decline in the size of the agricultural work
force has been greater than that of farm numbers and
that the number of males employed per farm has
fallen. On the adjusted figures of agricultural
establishments prepared by Lette (12), the number of
males employed per establishment fell from 1.8 in
1966 to 1.6 in 1971 and 1.4 in 1976.
     Since the demand for agricultural labor is
especially wage sensitive, the most influential
factor affecting this trend, which is due primarily
to the reduction in hired employees, was the rise in
nonfarm wages, with consequent increases in farm
wages. (13) A distinctive feature of the Australian
rural sector is in fact the close similarity, in
terms of both average income levels and the distri-
butional pattern of incomes, between primary producers and self-employed persons in other sectors;

this feature contrasts markedly with the experience in most other countries, including those of North America, where the income disparity between the rural and nonrural sectors constitutes a major rural problem. (14) As a partial consequence, no significant relationship exists between town size in Australia and the cost of food, clothing, and other main items of consumer expenditure aside from housing. Nor is there a significant relationship between the distance of a country town from a metropolis and the costs of consumer goods; only the cost of housing and services correlates with town size. (15)

In the postwar years until the mid-seventies, the movement of labor out of agriculture was facilitated by near-full employment in a buoyant economy, but the current depressed state of the economy with high unemployment tends to discourage outward movement. In any event, there would seem to remain a limited scope for further reduction in labor input per farm.

Accompanying the movement of labor out of agriculture were important changes in the age and sex structure of the agricultural work force. Since 1954 the male work force has been aging steadily, with the proportion of male workers aged fifteen to thirty-nine declining from 49 to 47 percent in the study period. Some evidence exists that structural adjustment in Australian agriculture is hampered by older farmers being less prepared to move away from farming when incomes are low than are younger farmers. (16) But the dependents of farmers with small holdings are more prone on reaching working age to leave farming than are the dependents of farmers with larger properties, and this tendency contributes to the trend toward larger farms. (17) Reliable statistics on the number of women engaged in agriculture are not available, but it is probable, given the substantial increase in husband-wife partnerships among farm businesses and the growing participation of women in nonfarm rural employment, that the number of women resident on farms who are actively engaged in agriculture has also increased. This probability is supported by the quarterly labor-force survey undertaken by the Australian Bureau of Statistics and obtained by personal interview: the survey indicates that female employment in agriculture rose from 61,800, or 15 percent of total agricultural employment, in May 1971 to 67,800, or 19 percent, in May 1976. The March 1976 agricultural census even puts female employment on rural holdings at 80,300 out of 336,300, or 24 percent, and the June 1976 population census puts it at 128,000 out of 385,000, or no less than 33 percent. Clearly, female labor would seem to be on the increase and a significant component of the total agricultural work force.

Since 1954 the rural population, which includes not only farm residents but also the residents of rural service centers with less than 1,000 inhabitants, has remained virtually constant, but relative to the national population, it has fallen sharply. In each of the census years 1954, 1961, and 1966, the rural population totaled 1.9 million; in 1971 it was 1.8 million, and in 1976 it was again 1.9 million; the difference between the 1954 and 1976 totals was a decline of less than 5,000. This remarkable stability highlights the shift from agricultural to secondary and, more particularly, tertiary employment within rural areas, a consequence of the substitution of capital for labor in the farming sector, of higher agriculture productivity, and of the need for more workers in country towns to service the requirements of capital-intensive farming. In each of the intercensal periods 1966-1971 and 1971-1976, almost as many people moved from urban to rural areas as moved from rural to urban areas; the totals for the 1971-1976 period were 237,000 urban to rural and 196,000 rural to urban. Yet the rural population expressed as a proportion of total population fell from 21 percent of a national total of 9 million in 1954 (the ratio had been 31 percent of 7.6 million in 1947) to 17 percent in 1966 (11.6 million) and 14 percent in both 1971 and 1976 (12.8 and 13.6 million, respectively). Over this period, even though the number of women in the rural population has remained fairly constant, the proportion gainfully employed has recently undergone a marked increase, having doubled from 11 percent in 1954.

The impressive stability in the national totals of rural population conceals considerable changes in spatial distribution. In the study period 1966-1971, the rural population grew most rapidly in the pastoral regions of the north and center as well as in the recently settled areas of western Australia, though in the Pilbara and the Northern Territory the pattern is somewhat distorted by mineral developments (see figure 5.5). On the other hand, almost all regions of intensive and semi-intensive agriculture, together with the remaining areas of extensive livestock grazing, experienced a net population

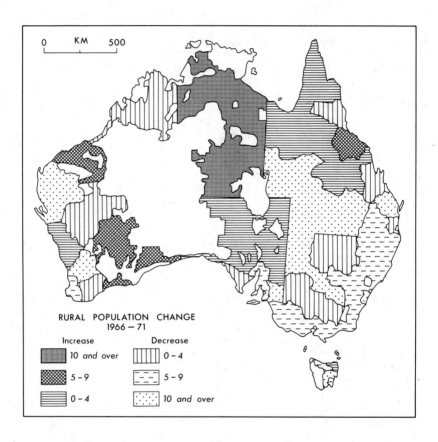

Figure 5.5.   Changes in the Size of Rural Population
in Australia.

loss. In the higher-rainfall coastal zones the
extent of loss was lower, despite lower farm productivity, than in the remote semi-arid tracts of
southeastern Queensland and northwestern New South
Wales, largely because opportunities for off-farm
employment are much greater in the coastal belt.
Thus, in 1975-1976 the proportion of properties with
household members working off the farm ranged from
about 50 percent in the coastal belt down to about
25 percent in the zone of extensive livestock
grazing. (18) Since Australia is a highly urbanized
country with its urban population strongly localized
within few centers, opportunities for off-farm employment are much less decentralized than in Europe

and North America. (19) Moreover, farm-labor requirements in the predominantly multiproduct livestock systems have less seasonal variation than in
cropping systems, though the seasonal nature of
labor requirements in wheat-sheep farming has led to
many farmers in the crop-livestock belts taking up
residence and seasonal employment in country towns.
(2) Consequently, much of the crop-livestock belt
shows only small increases or decreases in rural
population over the study period.

RURAL STABILITY

It remains to attempt a composite assessment of the
spatial pattern of rural stability in Australia. At
first glance, the pattern of rural population change
would seem to bear a much closer resemblance to that
of change in farm numbers than to the pattern of
change in the enterprise structure of farms (compare
figure 5.5 with figures 5.3 and 5.4). Yet, because
population change derives from a balance between
natural increase and the equilibrating mechanism of
internal migration, its pattern reflects in part
selective influences of spatial changes in farming
types. Accordingly, a single crude index of rural
stability was compiled by ranking the statistical
division data for each variable by rate of change,
by summing the ranks of the three variables for each
statistical division, ranking the totals thus obtained, graphing the totals, delimiting three arbi-
trary groups, and mapping the groups. (21)
     Although this arbitrary technique has obvious
limitations and the resultant pattern must be inter-
preted with caution, figure 5.6 nevertheless summa-
rizes the widespread stability that characterized
much of rural Australia throughout the most turbulent quinquennial period of the last quartercentury.
Stability appears to have been most pronounced in
two major regions: in western Australia, extending
from the northern sections of the crop-livestock
zone northeastward to the Kimberleys; and in eastern
Australia, extending from the Riverina northward to
the Cape York Peninsula. A minor region is coastal
South Australia. Since it would be necessary to
draw on much of the preceding discussion to inter-
pret in detail these seemingly highly stable regions, comment is confined to the inclusion of some
apparently anomalous areas that had a high rate of
change in rural population, farm numbers, or farming

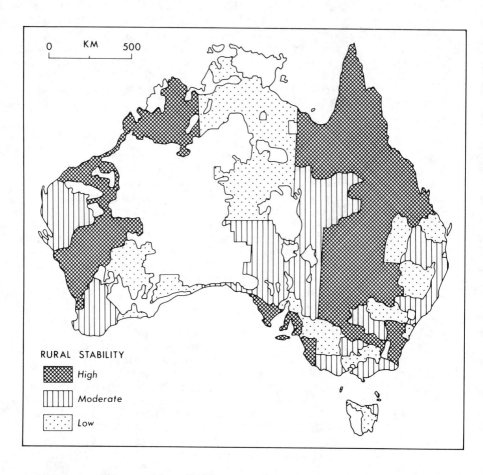

types. Thus, the Pilbara, despite a high rate of population increase due to mineral developments, enjoyed little change in either farm numbers or farming types; the Cape York Peninsula, despite some smallholding development, displayed scant change either in farming type or rural population; and northwestern New South Wales, despite substantial shifts in the enterprise structure of its predominantly crop-livestock systems and the establishment of irrigated cotton growing, continued to lose population at a relatively slow rate and showed only a small increase in farm numbers.

Instability appears to have been most pronounced in ten regions widely dispersed across the continent and less pronounced in eight regions largely confined to southern Australia. All regions except two underwent structural adjustments of their agricultural systems. But a striking difference exists between the fundamental changes that took place in the intensively farmed, southern coastal belts (where many farmers moved out of dairying or horticulture, and farms were amalgamated or abandoned) and the drier inland areas, where wheat-sheep farmers or graziers switched enterprises, and farm loss was much less. Thus, for instance, in the Maryborough district of Queensland, where tropical dairying had been very marginal, structural adjustment resulted in the loss of nearly a thousand dairy farms, one-seventh of the district farm total in 1966, and the establishment of some five hundred new and larger beef properties. Consequently, the district experienced not only a high level of farming-type change and a sharp fall in farm numbers, but also a substantial outmigration of farm workers and their families. Similar adjustments took place in the higher-rainfall regions of Tasmania, which had been dependent primarily on dairying or horticulture, and, to a lesser extent, throughout the coastlands from Brisbane to Melbourne and south of Perth. By contrast, the Roma district of Queensland merely underwent a shift from sheep to beef cattle; though this change was widespread, it reduced significantly labor requirements and resulted in a strong rural exodus. In the Wimmera and Mallee districts of the crop-livestock zone, the increased emphasis on sheep at the expense of wheat was accompanied by the amalgamation or abandonment of holdings and a moderate loss of population. Only two regions depicted as having low stability exhibited growth rather than decline: the Northern Territory with its expanding beef industry, and the Esperance sand plain with its large-scale land development program.

CONCLUSION

Over the past two decades, Australian agriculture has experienced a major shift in market orientation, substantial changes in farm enterprise structure, a decrease in farm numbers, an increase in average

farm size, marked changes in farm business organization, and a reduction in farm labor. Yet, even in
the period of most rapid and most pervasive change,
farm productivity continued to rise, and the loss of
farm population was, for the most part, balanced by
population gains in the rural nonfarm sector. In
many respects, notably its low fertility, low mor-
tality, high mobility, and semi-stable settlement
pattern, Australia's population exemplifies a
post-transitional society approaching the end of
demographic modernization. (22) Certainly, in the
turbulent late sixties, the rural sector displayed
outstanding stability, notwithstanding some structural adjustment, mainly within the more intensively
farmed coastal zone. In the future, such stability
can be maintained only through the continued evolution of highly flexible, capital-intensive, multiproduct pastoral systems that can accommodate considerable changes in the cost-price relationships of
their products and still achieve through scale
economies sufficient gains in productivity to guarantee the quality of life of the rural population.

NOTES

1. I.R. Wills, "Part-time Farming in Central Victoria," Review of Marketing and Agricultural Economics, 46 (1978): 196-219.

2. D.T. Rowland, Internal Migration in Australia,
Census Monograph Series (Canberra, Australian Bureau of Statistics, 1979).

3. D.T. Rowland, "Internal Migration as an Ex-
change Process: A Study of Victoria," Australian
Geographical Studies, 16 (1978): 15-28.

4. G. Lette, "Number of Farms and Farm Work-force
in Australia," Quarterly Review of the Rural Economy, 1 (1979): 147-150.

5. Ibid.

6. J. Dahlke, "Evolution of the Wheat Belt in
Westen Australia," Australian Geographer, 13: 14.

7. Bureau of Agricultural Economics, " The Economics of Land Development on the Esperance Sand Plain, Western Australia," Wool Economic Research Report, No. 17 (Canberra, 1970).

8. Wills, "Part-time Farming in Central Victoria."

9. Australian Bureau of Statistics, Agricultural Sector, Part 1 - Structure of Operating Units, 1975-76 (Canberra, 1977): 23.

10. J.M. Wells and W.R. Bates, "Changes in Farm Business Organization in Australia," Quarterly Review of Agricultural Economics 22 (1969): 53-65.

11. R. Smith and D.L. Smith, "Farm Fragmentation on Western Eyre Peninsula, South Australia," Australian Geographical Studies, 15 (1977): 157-173.

12. Lette, "Number of Farms and Farm Work-force," 149.

13. U.N. Bhati, "An Analysis of Aggregate Labour Demand and Supply Relationships for Australian Agriculture," Quarterly Review of Agricultural Economics, 31 (1978): 106-123.

14. E.S. Hoffman and J.R. Hume, "Farm and Nonfarm Income in Australia," Quarterly Review of Agricultural Economics 18 (1965): 121-135.

15. B.R. Davidson and M.F. Tierman, "A Note on the Comparative Cost of Living in Sydney and Country Towns in New South Wales," Review of Marketing and Agricultural Economics 43 (1975): 126-139.

16. K.W. Kerridge, "Value Orientations and Farmer Behaviour - An Exploratory Study," Quarterly Review of Agricultural Economics 31 (1978): 61-72.

17. P.W. Salmon, R.N. Fountain and H.S. Hawkins, Human Adjustment in Australian Agriculture 1972 (Melbourne: School of Agriculture and Forestry, University of Melbourne, 1973): 50.

18. P.C. Riethmuller and N.D. Spillman, "Off-farm Employment in the Grazing Industry," Quarterly Review of the Rural Economy (1978 Introductory issue): 44-47.

19. Wills, "Part-time Farming in Central Victoria," 220.

20. M. Williams, "Town-farming in the Mallee Lands of South Australia and Victoria," Australian Geographical Studies, 8 (1970): 179.

21. I. Pool, "A Method for the Social Grading of Areas," Pacific Viewpoint, 1 (1960): 225-237.

22. Rowland, Internal Migration in Australia, 173.

II

Modernizing Agriculture and the Rural Settlement Pattern

6 Environment, Land Use, and Development: The Case of Vermont

Gyorgy Olgyay

Vermont is a small state in the northeast corner of the United States. It is one of the three Northern New England states, (1) bounded by the Canadian border on the north, the Connecticut River on the east, by Massachusetts in the south, and by the long stretch of Lake Champlain in the west. (2) The state is classified by the United States Census as more "rural" than any other state in the Union except Alaska. (3) The rural quality of the state is enhanced by extraordinary natural beauty. Lakes, streams, and the wooded slopes of the Green Mountains create a natural and aesthetic environment of unique quality. This environment is the basis of the state's year-round vacation and tourism potential.

Vermont is not only rural, but also intensely agrarian. This is so in spite of the moderate-to-poor agricultural potential of the state in terms of topography, soil, and climate. The best agricultural areas are to be found in the river valleys, in the northeast corner of the state, and in the Champlain valley. (4) Today about 40 percent of all land under farm ownership is harvested or pastured cropland, representing about 20 percent of the state's total land area. Another 40 percent of farm acreage is wooded land, often on hillsides adjacent to croplands, contributing to the conservation of cultivated areas, watersheds, and drainage areas. (5) Since the post-Civil War period, dairy farming has been the state's primary agricultural activity. (6) The average annual milk production exceeds two billion pounds, supplying the

New England area with 48 percent of its dairy
product needs. About 77 percent of the state's
farm income comes from the sale of milk. (7)
During the 1960s, Vermont supported about 6,000
herds of dairy cattle, with 230,000 milking on the
average. (8).

Given the relatively "hard" natural conditions, Vermont's farmers operate on a high level of
technological efficiency. The farmer, be he dairy-
man, sheep raiser, or owner of apple orchards, has
to welcome technology in a drive for improvement or
perfection if he wants to be successful functional-
ly and economically. From its inception, this
piece of rural America tried to escape "imperfec-
tions" by inventiveness and intense utilization of
prevailing technology. For this reason, the state
has been called a "rural technopolity." (9)

The agrarian orientation of Vermont's rurality
has been reinforced and supplemented in recent
decades by the development of industries rooted in
the state's natural, scenic resources. The aesthetic quality of the environment supports all-sea-
son recreation, sport, and tourist activities. By
the beginning of the present decade, income from
recreation/tourism exceeded farm income by 35.9
million dollars. (10) While this figure indicates
a decline in farm acreage, farming and environment
based industries balance each other in a healthy
manner and represent a complimentary form of land
utilization. (11)

In recent years the balanced and developed
rurality of Vermont became threatened by the intrusion of nonrural elements. The threat did not take
the form of rapid industrialization, but the penetration of "suburban" development and life-styles
in their most disagreeable form. Vermont sensed
the danger of becoming an appendix of the megalop-
olis of the eastern seacoast. (12) In 1972 the
governor of the state announced that "Vermont is
not for sale," thereby symbolically summing up a
complex set of responses to a serious challenge to
the state's land-use profile. (13) The threat, the
challenge, consists of putting land into real
estate development, subdividing into vacation,
"second home" plots, and building suburban shopping
centers and malls. This trend was greatly facilitated by improved accessibility due to highway
building programs. The transportation infrastructure designed to link farm and recreation areas to

urban centers became the magnet drawing suburban
sprawl or splatter to the Green Mountains, affecting some of the most productive land areas. (14)
The land loss affected farms most severely, and the
process continues. It has been estimated that
Vermont is losing two-hundred acres of farm land a
day. (15) In central Vermont, the Regional Planning Commission estimates that the region lost
one-third of its farms during this decade. (16)

The threat to Vermont's farms, especially the
family farms, originates from several sources.
Some have to do simply with farm economy, such as
the need for high capital investment to maintain
effective technology, indebtedness, relatively low
return, and market conditions. The greatest danger
to the farm, however, is rooted in the impact of
high demand on real estate price structures, and,
consequently, on tax structures. Real estate is
subject to local taxation, on the basis of "fair
market value," not function or yield. The pressure
of demand for land threw the market value of farm
acreage out of balance. By taking a large tract of
land out of farm production and making it a
commodity, the developer changes the land's value
and places extra burden on the remaining farms. To
put it simply, the working farmer is squeezed by
skyrocketing real estate prices and taxes, and he
can ill afford to keep his land in cultivation.
When he sells, the land will not stay in farming;
it ceases to be a factor of production and becomes
a commodity. As such, land produces revenues in
taxes higher than that which farm products could
support. (17) Market values of farm real estate
are rising in the state at a rate greater than
anywhere else in the northeast region and greater
than the national average. (18)

This pressure on farm land also represents a
pressure on and a threat to the environment. It has
triggered defensive moves on the part of the state
in three forms: control of environmental permits
for development; statewide land capability and land
use planning; and tax policies.

Of these measures, the most important one is
the Environmental Control Act of 1970 (Act 250)
addressing both the issue of permit and that of
statewide planning. The law sets up the machinery
for the control of land use in terms of environmental criteria for all real estate development
over a certain size, 10 acres, for subdivisions of

more than ten lots, or the construction of more
than ten housing units. (19) All projects falling
into the designated categories must apply for
permit of land use to one of the seven district
environmental control commissions. Members of these
commissions are appointed by the governor, and are
citizens who hold no other state or local offices.
Each commissions' jurisdiction covers about
thirty-five town governments. The commissions conduct inquiries and hearings on pending applications, and their decisions to grant or deny permits
are final, with the possibility of appeal to the
state environmental board and the state supreme
court. The process followed by the commissions is
a quasi-judicial, adversary process to which the
principal parties are the petitioner and the local
community. All other interested parties, however,
are admitted to the inquiry, and the law emphasizes
that environmental and land use concerns properly
transcend the boundaries of immediacy. (20) Thus,
the regional impact of development and land use
changes is underlined and the multiplicity as well
as diversity of interests is recognized in applying
the criteria of the law.

Before granting a permit, the commissions must
find that the petitioner's project:

1. Will not result in undue water or air
pollution

2. Does have sufficient water available for
reasonably foreseeable needs

3. Will not cause unreasonable burden on
existing water supply, if one is to be utilized

4. Will not cause unreasonable soil erosion
or reduction in the capacity of the land to hold
water

5. Will not cause unreasonable highway con-
gestion or unsafe conditions

6. Will not cause an unreasonable burden on
the ability of a municipality to provide educational services

7. Will not place an unreasonable burden on
the ability of local government to provide govern-

mental services

    8. Will not have an undue adverse effect on the scenic or natural beauty of the area, historic sites, or rare and irreplaceable natural areas

    9. Is in conformance with statewide capability and development plans and land-use plans provided for by Law 250

    10. Is in conformance with duly adopted local and regional plans. (21)

    Implementation of the law has improved and slowed down real estate development in Vermont and has had a decidedly beneficial impact on the environment. Only a small number of applications have been rejected, but most permits have been issued with stringent conditions attached to them. Thus, the operation of the law has a wider and more pervasive effect than the actual negation of permits would indicate. (22) The law has provided Vermont with nine years of experience in land-use control, an experience which is evolving and subject to legislative scrutiny as well as continuous and intense public debate. This experience highlights the public concern for a given cherished rural life-style and the environmental preconditions for it. This experience also illustrates in land-use cases that, inevitably, public and private rights and interests clash, and the quasi-judicial process of adversary proceedings must be looked upon as a form of conflict resolution. (23)

    The multilayered conflict of rights and interests was well illustrated by the most celebrated case decided under Law 250. In 1976, the Pyramid Corporation of Dewitt, New York, a regional developer of shopping malls, initiated a two-year process aimed at the construction of an eighty-two store mall in the town of Williston, not far from the largest city of the state, Burlington. (24) On October 12, 1978, the Fourth District Environmental Commission denied the necessary land-use permit on the basis of highway congestion, undue burden on both public and private services, and nonconformance with local and regional plans. (25) The Pyramid project generated heated discussions. The commission held thirty-eight hearings in the case, admitting as parties practically all the adjacent

municipalities, three regional planning commissions, and diverse private interests. (26) Although pro-mall arguments pointed to benefits to be
realized by Williston in the form of a significantly enlarged tax base and new employment opportunities, opposition to the project developed even
within the city on the basis of long-range considerations, environmental fears, and the regionally
adverse impact of such a large commercial project.
(27) In addition, fierce opposition to the mall
emerged in Burlington and on the part of the
regional planning authorities, as the mall represented a threat to the revitalization of downtown
Burlington as the core of both the city and the
county. (28) All surrounding communities feared
the adverse impact of an explosive increase in
vehicular traffic both in settled areas and on
highways not designed for such schemes.

In the debates we can observe the conflict of
rights and legitimate interests, both public and
private. For example, in the conflict of municipalities, Williston and Burlington, we must recognize that Burlington's well-being as a community
and administrative/political entity depends to a
large extent on the city's role as a commercial and
social/cultural core of the region. The disruption
of that role was threatened by Williston's "good
fortune," the mall, and neither Burlington nor the
region could remain indifferent to the threat. The
term "environment" quite clearly involves more than
physical surroundings, and concern for the health
of the social environment also has its impact on
the operation of Law 250.

Our discussion so far has dealt with only one
of Vermont's three forms of attack on land-use
problems. Admittedly, the control and permit is-
suing process is the most developed part, but brief
mention must be made of land capability and
land-use planning and of tax policies.

As we have seen, the operation of the commissions is regulatory rather than indicative. Therefore, the legislature under the same Law 250
mandated the state environmental board to develop a
series of land-capability evaluations and land-use
plans. Such plans should identify and indicate the
locus and type of development that is desirable and
worth encouragement. This process is statewide,
comprehensive, and controversial. The capability
studies and plans provide guidelines for a state-

wide system of "zoning." Many issues, however, are
still left unresolved, such as the relationship of
state and local zoning authorities, or what constitutes desirable develoment in terms of long-range
considerations. (29) Clearly, however, the cou-
pling of environmental control and land-use planning points the way for the state.

The third category of Vermont's measures addressing the problems of land use is tax policies.
One such tax measure is a scheme of property tax
credit tying the real estate taxes of state residents to income and placing limits on taxes to be
paid, 4 percent and 6 percent depending on income.
By limiting the credit to persons living in Vermont
as permanent residents, the scheme clearly benefits
Vermonters and discriminates against the beneficiaries of "suburban sprawl," the vacation home
owners. (30) In addition, tax policies also involve a special capital gains tax imposed on quick
profits gained by rapid real estate sales. This
policy taxes heavily those real estate transfers
which take place within the time frame of six
years. The more rapid the transfer and the greater
the profit, the heavier the tax burden. (31)

Both of these tax measures are designed to
slow down the loss of farm land to developers and
to prevent the rapid transformation of land from
factors of production to commodity. Both of these
measures have achieved results, but they do not
address the problem of the heavy tax burden of
farmers, rooted in the principle of taxing land
based on "fair market value." This particular
feature of the tax squeeze had a debilitating
effect on the family farm. Proposals were under
discussion which would radically alter this principle by taxing land on the basis of use and
function, separating farm land from other forms of
real estate. Such policies were to provide much
needed relief for crop and timber land, lessen the
tax burden on the family farm, and slow down even
more the sell-off of prime agricultural land to
real estate developers. In response to such considerations, a tax measure based on land use was
passed in 1978. The impact of the new law, however, cannot be estimated as of now. (32)

Some lessons may be drawn from the preceding
survey:

1. Vermont, as a rural community, is con-

cerned about the pressure on physical, aesthetic, social, and economic environment flowing from changes in land use due to development.

2. The state's concern has brought forth measures designed to slow down and direct the process of change. These measures include regulation and control, guidelines and plans, tax policies and tax reforms.

3. Such measures are implemented on both the state and the local level. They are fluid, multifaceted, and subject to constant public scrutiny resulting in reform and revision. Controls are flexible, decentralized, and quasi-judicial, not rigid, centralized, and administrative.

4. The state is determined to protect not only a given socioeconomic profile and structure, but also a unique life-style. One may argue that this determination is conditioned by Vermont's historical "ideology" of rural populism.

5. The policies of the state can be considered as trend setters for other regions of the nation. It appears that changes in land use has slowed down and has improved considerably because of the operation of the system.

NOTES

1. The three Northern New England states are Vermont, New Hampshire, and Maine. Vermont's total land area is 9,609 square miles and the population according to the 1970 census was 444,732.

2. The boundaries of Vermont were defined in history through conflicts with her neighbors. Apart from Texas, Vermont was the only state that existed as an independent republic prior to her joining the union as the fourteenth state (1777-1792).

3. n locating a given state on the rural-urban continuum, the Census Bureau uses population density as the sole criterion. This method, of course, does not take into consideration occupation, function, or infrastructure, and does not address itself to the question of rural-urban

life-style or consciousness. See Frank M. Bryan, Yankee Politics in Rural Vermont (Lebanon, NH: University Press of New England, 1975): 241 ff.

4. Harold A. Meeks, The Geographic Regions of Vermont: A Study in Maps (Lebanon, NH: University Press of New England, 1975): 68-76. Also, Benjamin Huffman, The Vermont Farm (Montpelier, VT: Vermont State Planning Office, 1973): 3.

5. Huffman, 3.

6. Early, pioneer, subsistance farming was followed by sheep raising as a specialized form of agricultural activity during the early part of the nineteenth century. The dairy cattle replaced the sheep during the second half of the century.

7. Huffman, 4.

8. Bryan, 243.

9. Ibid., 234-267.

10. For 1971 the corresponding figures were: farm income $170.3 million, income from recreation/tourism $206.2 million. Cf. Huffman, 65.

11. In 1949 approximately 60 percent of the state's land was under farm ownership. Today about 20 percent is. Cf. Huffman, 39-41.

12. Robert G. Healy, Land Use and the States (Baltimore, MD: John Hopkins University Press, 1976): 35.

13. Bryan, 253.

14. John Marshall, Jr., The Efficacy of Act 250: The Evolution of an Environmental Law (Montpelier, VT: Environmental Planning Information Center, May 1971). See also Biennial Report of the Commissioner of Taxes, 1972 (Montpelier, VT: Vermont Agency of Administration, 1972). Seasonal homes were estimated to be worth $459 million in 1971, an increase of 114 percent in four years.

15. Burlington Free Press, July 7, 1979.

16. Burlington Free Press, July 10, 1979.

17. Healy, 55.

18. Huffman, 18.

19. 2.47 acres equal 1 hectare. All construction, regardless of size, at altitutdes higher than 2,500 feet is subject to regulation. See Healy, 38.

20. Healy, 39. See also Vermont's Land Use and Development Law, Environmental Board Regulations Annotated (Montpelier, VT: State Planning Office, 1976): 19-20.

21. Ibid., 21-26.

22. Healy, 40.

23. This is especially worth emphasizing in view of the fact that developers, in anticipation of objections, conditions, or adverse ruling, will tailor their projects and plans in such a way as to minimize the resistance to them. Thereby the level of potential conflict is reduced.

24. Both municipalities are in Chittenden County. Williston's population is about 3,000, while Burlington's is over 40,000.

25. State of Vermont, District Environmental Commission #4, Order Denying Land Use Permit, Application #4C0281, p.3.

26. Ibid., B-1.

27. Burlington Free Press, September 21 and 22, 1977.

28. See statement of Burlington's Mayor, Gordon Paquette, on June 17, 1977, in Burlington Free Press. Also, Chittenden County Regional Planning Commission, Memorandum: To District Environmental Commission #4, May 15, 1978.

29. Cf. Vermont Agency of Environmental Conservation, Environmental Board, Interim Land Capability Plan, Montpelier, VT, 1972. See also Healy, 50-52.

30.   Healy, 57.

31.   Ibid., 58-59.

32.   Huffman, 22-27. The erosion of farm property, the loss of farm land to real estate development, is not only a Vermont, but also a national, phenomenon. In view of worldwide food shortages, it could potentially develop into an extremely serious situation, the significance of which tran-scends the boundaries of Vermont. For evaluation of the 1978 law, see Burlington Free Press, October 18, 1979.

7 Farming Space and Migration: The Case of France

Violette Rey

INTRODUCTION

Modern agriculture can be defined by a new combination of its three factors of production, in which capital, in the form of materials of various technical and financial capacities, has acquired a determining position over the other factors, namely, the land and the people. Besides, within the different sizes peculiar to each system of production, the accountable profit favors those farms where the man/land ratio is lower because the income proves to be larger. The general impact engendered by that double characteristic of modern agriculture has made itself felt through the elimination of many farms unfit to follow suit, to the benefit of a continual enlargement of the remaining ones. That mechanism maintains a constant pressure on the enlargement which generates new displacements among the farmers; the latter will seek further for larger lots to till, and, from a distance, cause new tensions over the settling spaces.

Such displacements, hard to detect because neither the farming institutions nor the farmers of the immigration area consider them favorably, seem to progress and perhaps denote the beginning of a movement tending to protract for a long period. Even if they cannot always be easily dissociated from the traditional farming currents of emigration and if they represent minor fluxes in work power, they differ by their means and disclose what stakes are competing to gain control over the farming area. Their impact upon the rural surroundings is generally very strong, but often controverted.

NEW AREAS OF DEPARTURE/NEW AREAS OF ARRIVAL

The same recent tendency appears at three different geographical levels: the initiative of the migration is taken up by the farmers of the rich agricultural regions, who turn toward lands whose enhancement is technologically less intensive. This is also true of the farm people shifting about in France, of Europeans settling in France, and of the beginnings of French farming emigration across the Atlantic.

Migratory Shiftings in France since 1950

Important even before the war, the migratory movement of the farm people has been helped since 1949 by the Ministry of Agriculture, which enables one to better appreciate its importance. At that time, France was divided into two areas, an area of departure north of a line stretching between La Rochelle and Les Vosages and an area for migrants in the south; the division was established in reference to the supply and demand of farms, to create a consistent farming market rather than balance agricultural densities. Between 1949 and 1973, a period of considerable displacement, about eleven thousand families (fifty thousand people) left to settle in the French immigration area, leaving behind 140,000 hectares in their parent lands to take possession of 430,000 hectares in the immigration area.

The bulk of the movement was before 1965; since 1973, because of significant reduction in governmental assistance, no applications for resettlement have been received. For the whole period, the area of massive departures are the Breton, Norman, and Picard borders, and all the Aquitainian southwest is a huge immigration area. However, a chronological division emphasizes the radical change in the geographical origin of the migrants: up to 1955, the west supplied the main effort (68 percent of the national movement); during the period 1960-1965, its contribution dropped (45 percent) to the benefit of the Parisian Basin and

the north (36 percent). During the 1970s, the west
was totally outdistanced (20 percent) by the intense cereal-growing regions of the Parisian Basin.
The same change can be noted in the areas of
arrival: the contact of the Parisian Basin with the
Massif Central is more important than that with
Aquitain. The exhaustion of official support and
the decline of registered applications wrongly led
some to believe that farming emigration in France
was ending. Indeed, it continued to decrease because the available lands had become scarce in the
immigration areas, and because the local farmers
absorbed the surplus directly; however, farming
emigration has not disappeared, but has undergone
some changes. On the one hand, the migrants of
today are few and settle over areas far larger than
those which their predecessors settled on (24
hectares in the decade of the fifties versus 80
hectares in the seventies). On the other hand, an
increasing number replaces the settling emigration
by the purchase of a second farm.

Renewal of Foreign Farmers Settling in France

The lack of balance in the density of the farming
populations and in the prices of the land and the
rents, which are definitely lower in France than in
neighboring countries, have for a long time favored
the infiltration of non-French farmers into French
lands. Two streams could be perceived before the
war: the farmhands and peasants immigrating from
Spain and Italy settled on the small farms deserted
by the agricultural exodus in the southeast and
southwest; the Belgians and Flemish again took
possession of the spaces broken up by the bombing
in Picardie, Artois, and Normandy. Under the
clauses of the Treaty of Rome (1957) concerning the
free circulation of people within the European
community, France obtained a lengthened program for
free settling. (1)
      The movement, however, has yielded few fruits.
In the 1960s, there were an average of 170 cases
involving 4,500 hectares per year; in the 1970s, it
amounted to only 150 cases and 3,200 hectares. But,
above all, the movement changed characteristics:
although the Italians ranked first and the Spanish
fourth, the neighboring farm people of the north
have become the majority; Belgians, Dutch, and
Germans progressively assume the larger portion of

those new settlements. The nationals of the E.E.C.
now take from 60 to 70 percent of the settling
areas cultivated by foreigners, as contrasted with
their 45 percent from 1962 to 1973. These figures
still better reveal the cleavage between the two
migrations when one considers the average size of
the farms recuperated by each nationality. The
Germans are foremost with 39 hectares, followed by
the Dutch (37 hectares), the Belgians (32 hectares), the Swiss (29 hectares), and the Danes (26
hectares) who started to immigrate in 1974. Con-
versely, the Italians, Spanish, and Portuguese
fluctuate between 12 and 18 hectares. The level of
development at the time of location confirms the
suspected cleavage: while the Mediterranean immigrants still chiefly decide in favor of rent and
sometimes even sharecropping, the immigrants of the
north buy their farms. Traditionally located in the
southwest and the south of France, the foreign
farmers have just started an enlargement of their
estates, particularly in the center, where in three
years 1,500 hectares have been occupied again by
Germans, Belgians, and Swiss. The locations of the
settlements have nothing to do with proximity to
borders; the location of farming space that is vacant is the foremost factor of relocation.

              Resumption of French Farming Emigration
                      Across the Atlantic

There had still been a certain flux of location of
European farming colonies from across the ocean in
the 1950s. But in the 1960s, still scarred by the
decolonization and the flowing back of farm people
to the metropolis (pieds noirs from North Africa
who settled again in the south and southwest of
France), struck also by the European economic
expansion and the very strong attraction felt by
the surplus farmhands towards the urban system,
there was a very limited movement of agricultural
emigration. Yet, since the 1972-1975 period the
influx of European migrant farmers across the
Atlantic has started again, but with new modifications: individual initiatives from the migrants and
greater requirements from the local immigration
areas regarding technical means and financial self-
support of the newcomers. The emigrants almost
exclusively go to the Americas: the United States,
Canada, Brazil, and especially Argentina. Knowing

that in the Americas one does not incur political
risks comparable to those in Africa and that one
may still find enormous vacant spaces, a few French
farmers, like other European farmers, have once
again departed, though no one seems to know how
many have left since 1976.

An exhaustive survey of this movement is
extremely difficult; the information given is limited to the statistics kept by the Federation des
Associations de Ruraux Migrants a l'Etranger
(FARME), the chief information organization helping
farmers who wish to emigrate, which was created in
1976. Through analysis of the distribution of its
members in the departments and a survey of sub-
scribers to its review and the people joining in
the meetings, an approximate map of the degree of
interest (if not an actual application) could be
established. It is quite likely that if that
organization and the movement it helps were better
known, the actual application would be greater and
more widely distributed. However, this potential
movement of foreign emigration generally follows
the farmers' migration within France. The largest
number of people interested are the farmers of the
Parisian Basin, especially the Picards and Marnais,
who, in a system of vast cultivation, are most
interested in undertaking the greatest initiative
in the matter of intensive farming combines with
agronomy and nutrition.

With southern France still an open area, the
Aquitanian departments - Bouches-de-Rhone and Allier - share in the movement. In most of those
departments, there are two cumulative factors of
mobility: (a) the previous locations of migrants
(metropolitan, repatriated, or foreign) have introduced an agricultural mentality open to mobility;
(b) subsequent natural calamities have heaped difficulties on those who were further engaged in
modernization and were most indebted, often to
farmer migrants. But why were the hills of the Gers
or the sands of the Allier so easy for newcomers to
take? Was it not partly because of their irregular
crops? Thus, the fundamental characteristic of
contemporary farming migration is that it is a
marginal movement with a high risk of uncertainty.

The Migration Quickens the Agricultural
Modernization and the Rural Incorporation
in the Immigration Area

The multifarious effects in agricultural innovation
and diffusion are sufficiently known to mention
them only briefly: the improvements of the soils
(drainage, irrigation); the extension of cultivated
land (over fallow land or by deforestation); the
changes in the methods of farming and in the
diversification and increase of produce; the development of the supplying areas; and commercialization and technical staffing.

After some time of critical observation, the
neighboring farmers imitate the newcomers and join
in the process, transforming their production system, then their ways of life and their prospective
relations. These dynamics of transformation are
explicitly sought and expected by the authorities
responsible for the rural development in Latin
America. The study of the effects of postwar
farming colonies shows to what extent they appear
to be among the best channels to convey the modern
techniques of cultivation and the complex aspects
of rural organizations.

Migration, however, is not automatically successful; its effects largely depend on the clever-
ness with which the migrants manage in their new
suroundings. Indeed, this contemporary migration
belongs to a dis-symetric process. The migrant
arrives better instructed, better equipped, richer
than the farmers among whom he settles; his aim is
not to adopt the local trends, but to enhance his
own techniques of production, those which are
familiar to him or new ones with which he hopes to
get rich. His activity then disturbs the previous
rural and agricultural organization and creates
tensions in his new location.

The example of innovation among the local
farmers sharpens the tensions between conservative
and modernizing tendencies. The migrants often
treat the countryside they have "conquered" roughly; they go on extending their area far, and soon
the rise in the price of land is such that it makes
it impossible for latecomers to purchase new acre-
ages. Linked to their needs in multiple services,
the initiative of the migrants goes beyond the
farming domain; after a few years, they number
among the trade-union leaders and sometimes beccome
mayors of their "communes". As a rule, it is among

the nonfarming country people or among distant
farmers that the migrants find support for their
incorporation and social rising, while the local
farmers endeavor to check their alien and competing
presence.

The failures of the immigrants are many. Among
the many factors likely to bring failure about is
the inhospitality of the immigrant environment,
particularly significant from a geographical point
of view.

Whatever his assets may be, the migrant farmer
who settles in the tail of a migrating chain is
compelled to settle on poor soil and in marginal
regions, in comparison to his original environment.
If he makes any error in the estimation of physical
conditions (infertility, a wrong estimate of the
water supply, a lack of balance in the funds
inconsiderately locked up in landed property and
its improvements, the expense of the production
equipment), he may go bankrupt or resort to uncer-
tain shifts to find some fresh supplies of money.
The immigrant may be forced to grow tobacco instead
of maize, or his lumbering may turn into charcoal.
These are just a few of the pitfalls. In other
respects, he is more likely to achieve success if
the immigrant environment offers some receptivity
or is similar to his original surrounding; in
particular, by having a wide structure of land
ownership and tolerant local people.

All things considered, does the contemporary
migration of rich farmers play a positive part? One
wonders about the still modern character of so-
called modern agriculture: it is still based upon
plentiful funds, cheap supply of power, and easy
gains in productive capacity. In fact, contemporary
farming emigration occurs only in regions that have
become marginal, are on the wane demographically
and economically, are poor, impoverished, or are
not yet enhanced. What would some areas in the
Limousin, the Gers, or Chaco be without the presence of migrants? They would be still more deserted. Besides, considering it impossible to find
only one solution to the problems of the distribution of resources and economical development, the
farmers' emigration appears to be a solution whose
effects have nothing to do with the farmers' work
power, but the great risk of the emigration must
always be emphasized.

NOTES

According to Decree No. 54-72 (January 20, 1954), any foreigner is allowed to settle as a cultivator on French soils if he: (a) has already worked two years in France as a paid farmhand or (b) settles on wasteland or on forest lumber lands. Any other settling requires a special permit from the Prefecture.

REFERENCES

Bonnamour, Jacqueline, "L'eclatement des exploitations," Bulletin de l'Association des Geographes Francais, No. 397-98, (1972): 131-134.

Ministere de l'Agriculture, "Installations et acquisitions des etrangers en France," Rapports 1973, 1974, 1975, and 1976.

Rey, Violette, "Le theme de la migration agricole en France," Cahiers de Fontenay, No. 7, (1977): 65-91.

Rey, Violette, "Les agriculteurs migrants: Le cas du Cher," Etudes Rurales, No. 68, (1977): 55-91.

8 Agriculture and the Rural Settlement Pattern: The Case of Finland

Uuno Varjo

A fundamental change has occurred in the status of agriculture in Finland over the last twenty years. While in 1960 a third of the country's population was engaged in agriculture, this proportion has now declined to approximately 13 percent, and a corresponding drop has been seen in the number of farms, from almost 400,000 in 1960 to about 250,000 in 1979; that is, a rate of 7,500 farm closures per year. The most pronounced alterations in the pattern of agriculture in northern Finland are those attributable to reduced productivity and the effects of the field reservation system, introduced in 1969 in response to overproduction in foodstuffs (1), as a result of which the number of farms in many communes fell immediately by one-fifth (the decline for the year 1978 over all of northern Finland being about 12 percent (2). It is only to be expected that changes of this magnitude will have left their mark on the attitudes of the local people toward both agriculture and rural life in general.

Research into the abandonment of agriculture has been carried out in Finland since the 1950s, when it was pointed out, for instance, that a zone had grown up around Turku, the largest city in southwestern Finland, in which agriculture was relatively extensive in character in spite of the favorable location. (3) An inquiry into the causes for this showed that, for a variety of reasons, the area was suffering from a shortage of labor. Perhaps the most important factor was the boom in industry following the Second World War, causing

industrial wages to rise more rapidly than agricultural productivity,. while industrial work was,
nevertheless, frequently easier, cleaner, and more
regular. Thus, where the two means of livelihood
were competing with each other for labor, this soon
led to a victory for the industrial sector, although this effect did not extend more than 30 or
40 kilometers out from the city.

FUTURE PROSPECTS FOR FARMERS IN LAPLAND

The situation is rather different today, however.
In Lapland, for example, the vigorous settlement
projects and economic growth of the early 1960s
have subsequently turned to stagnation and decline
in almost all fields. (4) We first set out to study
this aspect in 1969, by means of a questionnaire
sent to all the 17,000 farms functioning in the
province of Lapland at that time, asking for the
farmers' opinions on the future prospects for their
farms. From this it emerged that the whole province
was affected by a great uncertainty regarding the
continuation of farming: only 27 percent of the
respondents considered that the future of their
farms was secure. The greatest degree of uncertainty was experienced in the best agricultural area,
the southwest of the province. The people's faith
in the future remained strong in the more remote
districts of the far north, with over 60 percent of
the farmers in many villages, and over 80 percent
in some, being sure that there would be someone to
take over the farms from them. There does, in fact,
seem to be a clear positive correlation between
optimism concerning future prospects and distance
from the center (see Table 8.1); for, in the
individual communes, too, the farmer was aware of a
successor less often in the area surrounding the
commune center than was the case further away.
    The continuity of a farm is closely bound up
with its viability, which the farmers were also
asked to evaluate. Here, the same differences in
attitude between the regions within the province
emerged. Although those farms estimated to offer
the farmer a good living were likely to be farmed
by the following generation in an average of 40
percent of cases, a distinct difference is seen
between the figures for the southwest and those for
the more remote areas. The same was also true when
future prospects were examined against farm size:

there were some communes in the southwest of the province in which only 10 percent of the farms in the smallest size class were certain to be worked in the future, while over half of such farms in Enontekio and Utsjoki in the north were in this position. Although it is true that reindeer herding and incomes from outside farming serve to ensure an adequate livelihood in the latter case, (5) it was also clear from the 1969 statistics that the population in the north was more ready to accept a lower standard of living than that in the south-western part of the province.

Table 8.1. Correlation Between Distance
From the Nearest Center and Number of Farms
With a Known Successor

Distance (km)	Number of Farms	Percentage With Known Successor
less than 10	3398	23.6 percent
10-20	2724	27.8 percent
20-50	6029	29.5 percent
more than 50	1787	32.9 percent

FIELD RESERVATION AND ABANDONMENT OF FARMING

A similar pattern may also be observed in the incidence of field reservation and abandonment of farming. The field reservation system affords us one opportunity for examining the attitudes of farmers toward agriculture, since it allowed them to give up farming in return for compensation from the state. (6) The effects of this system over the period 1969-1971 have been studied in the case of Middle Bothnia comprising twenty-two communes and three towns. (7) The material included about four-teen hundred farms placed under such a reservation agreement. The mean distance of these from a center may be seen in Table 8.2. Whereas 36 percent of these farms were located closer than 5 kilometers to a center, only 8 percent were more than 15 kilometers away. Thus, it seems that the field reservation farms were fairly highly concentrated in the zones around the centers.

Table 8.2. Number of Farms Under Field Reservation
and Distance from Commune Center in Middle Bothnia
in 1971

Distance (km)	Farms (percent)
less than 2	15.9
2-5	20.6
5-10	32.9
10-15	22.8
more than 15	7.8

At the same time, the higher the proportion of
farms in the commune which are under such an
agreement, the greater the concentration. The weak-
est agglomeration of reservation farms was found in
Kaustinen, where the proportion of such farms was
low, while the most pronounced agglomeration was in
Lestijarvi, where the proportion was high. Thus,
field reservation in northern Finland is not an
exclusive characteristic of the remoter districts,
nor was it connected with the lack of services, or
the lack of an inspiring living environment, for
instance. Rather, it showed a marked concentration
in the areas around the centers. This is also
reflected in the fact that the influence of the
centers seems to have spread year by year by a
process of diffusion to gradually more remote
areas. (8) For example, 58 percent of the field
reservation agreements entered into in 1969 in the
commune of Ruukki, about one hundred kilometers
south of Oulu, involved farms around the three main
villages, from which the trend then spread to the
more sparsely populated parts of the commune over
the years 1969-1974 even though the most pronounced
centers of field reservation remained to some
extent unchanged. Even in these areas, however, the
trend has spread gradually towards the extremities
of the villages.
    It thus seems that agriculture in the zone
adjacent to the center has not been able to resist
the pressures exerted on it by the center itself,
and has consequently entered into a decline, whereas a greater confidence in future prospects has
prevailed in the more isolated villages.

SUBJECTIVE IMPRESSIONS OF THE FARMING VILLAGES

The healthy development of a farming village may be regarded as entailing a certain measure of attractiveness as a place to live, of activity on the part of the village community, and of continued social interaction among the inhabitants. These aspects were studied in 1977 by means of another questionnaire, administered to all farms in a total of nine villages in the remote country district of Kuhmo and the rural commune of Keminmaa. This involved people living permanently in these villages and receiving a steady income.

Kuhmo is a commune of almost fifty-five hundred square kilometers immediately adjacent to the Finland-Soviet Union border, and is of a size which would engulf about thirty of the communes of southwestern Finland. Six villages were studied, all of which lay at a distance of twenty-five to fifty kilometers from the commune center. Replies were received from about two hundred and ninety out of the total population of just over three hundred families. Keminmaa is located between the towns of Kemi and Tornio, and the three villages selected together possessed just under two hundred families, 140 of which were interviewed.

Only 3 percent of the inhabitants of the six villages in Kuhmo replied that they did not like living in their village, and only one person in the three villages in Keminmaa answered in this way. There are, of course, many factors implicated in the satisfaction of inhabitants with their environment, and it would be impossible to determine these other than through certain characteristics of the attitudes expressed. The method used for this purpose, which was apparently first developed for research into "company images" (9) is based on the analysis of "semantic differentials," in which the respondent is asked to choose between a pair of dialectically opposite concepts as representing his opinion and indicate the strength of his conviction by a score from 1 to 5, with the lowest score denoting the strongest negative conviction and the highest the strongest positive one. The conceptual opposites used in regard to each village were the following:

- Friendly - Unfriendly
- Rich - Poor
- Active - Passive

- Liberal - Restrictive
- Clean - Polluted
- Well supplied with Services - Poor in Services
- Modern - Old-fashioned
- Varied - Monotonous

The results for the individual villages in Kuhmo are shown in figure 8.1, in which the darkness of the squares in the columns indicates the weight of opinion in favor of that alternative, and the vertical line indicates the weighted mean value of the distribution of all replies. General features worth noting are that the villages all appear friendly and clean places to their inhabitants, and most of them are also modern places. Many other positive judgements are made, but there are also some negative ones, such as the low level of services and in some villages a complaint of monotony; that is, a lack of stimulus or inspiration.

Another way of measuring the acceptability of a person's home village is in terms of his desire to move away, and thus each respondent was also asked where he would best like to live, given that work and accomodation could be found there, the alternatives being his own village, the commune center, the nearest town (Kajaani), or some large city such as Helsinki. The results again pointed in the same direction, in that about 90 percent regarded their own village as still being the best place to live, with 8 to 9 percent preferring the local center, and a few individuals favoring a nearby town, and nobody wishing to move to a large city. The inhabitants, thus, do not seem to view their remote villages in northern Finland with disfavor; the villages are not driving them away.

SOCIAL INTERACTION IN THE VILLAGES

The functioning of a rural village and the creation of a pleasant environment in which to live require a sense of community in the village and the preservation of the system of social interaction within this community. The pressures being exerted upon the country districts in northern Finland are effecting a change in the functional nature of the villages in this respect, and this is particularly in evidence in the areas adjacent to the centers. The traditional basic source of livelihood in the

the local bus services of the latter, so that the
services of both these centers are readily available. Viitakoski, on the other hand, is a more
remote village lying 25 kilometers away from the
commune center and with poor connections to it.
This aspect of location with respect to a center is
clearly reflected in the occupational structure of
the villages, and above all in the position of
farming, which still occupies approximately 40
percent of the labor force in Viitakoski, but no
longer more than 10 percent in Liedakkala. A
corresponding difference is found in the service
occupations, which account for just over one-fifth
of the labor force in Viitakoski, but about one-
half in Liedakkala. Thus, Liedakkala may be said to
have become substantially urbanized by comparison
with Viitakoski, as may also be seen from the mean
age of the farmers (Table 8.3), around 60 years in
the former village, but only 42 in Viitakoski. It
does indeed seem that farming is dying out in
Liedakkala, as is also shown by the fact that no
successor was known who would take over any of the
farms, while 71 percent of the farmers in Viitakoski were sure that their work would be continued.

Table 8.3.   Comparison of the Villages
of Liedakkala and Viitakoski

	Liedakkala	Viitakoski
Mean farm size (arable area, hectares)	9.5 ha	9.4 ha
Successor known	0	71 percent
Mean age of Farmer	60 years	42 years
Number of contact persons in village	3.3	5.3
Desire for cooperation	33 percent	89 percent
Mean income per employed person	29,730 Fmk	20,245 Fmk

These two villages also differ in their income
statistics. The mean income per employed person in
Liedakkala in 1975 was about thirty thousand Finnish marks (Fmk), of which only 4,000 Fmk would be
from agriculture, whereas in Viitakoski the total
income per person was smaller, only 20,000 Fmk, but
the proportion gained from agriculture was very
much larger at about fourteen thousand Finnish
marks.

In spite of the fact that family incomes in
Liedakkala are very much higher, the availability
of services is better, and communications with the
nearby centers are good, social intercourse runs at
a very much poorer level than at Viitakoski, which
has remained a lively, functioning village in this
respect. It is obviously the case that the proximity of a center, allowing a greater diversity in
means of livelihood and spare-time interests, undermines agriculture and divorces the living environment of the individual from his daily work and
interests. In this way it serves to alienate the
inhabitants from the village community, that their
own mutual contacts are severed. Once this happens,
the village is no longer a living entity, but a
sterile dormitory community. In the remote farming
villages, on the other hand, life and work share
the same surroundings and the people also share the
same interests. This then creates contacts and
stimulates attempts at cooperation, which is the
basic requirement if the village is to live and its
community is to develop.

SOCIAL INTERACTION AND THE FUTURE PROSPECTS
OF THE VILLAGES

The above comparisons between the attitudes of the
inhabitants of certain villages in northern Finland
toward their surroundings, their neighbors, and
their future prospects serve to demonstrate the
extent to which the centers have reduced the
viability of agriculture in the adjacent areas. Our
conviction is that the interaction and future
prospects in the villages improve with increasing
remoteness from a center. This being the case, it
seems that growth-center policies, which favor the
centers exclusively and attempt to develop their
economic life, and then wait for the affluence
achieved by this to radiate outwards into the
surrounding countryside, have failed to achieve the
desired result. On the contrary, it would seem that
in northern Finland, at least, rural settlement has
retained its vigor best in those areas which lie
beyond the direct influence of any center.

NOTES

1. Uuno Varjo, Finnish Farming: Typology and Economics (Budapest: Akademiai Kiado, 1977): 28.

2. MR. No 7/1978. Monthly Review of Agricultural Statistics, No. 7 (Helsinki: Board of Agriculture, Statistical Office, 1978).

3. Uuno Varjo, "Zonengliederung in sudfinnischer Landschaft und Landwirtschaft," Fennia 82, No. 4 (1958): 73-74.

4. Uuno Varjo, "Development of Human Ecology in Lapland, Finland, after World War II," Geoforum 5/71 (1971: 47-74. See also Uuno Varjo, "Agriculture in North Lapland, Finland: Profitableness and Trends since World War II," Fennia 132 (1974): 36-66.

5. Eino Siuruainen, "The Population in the Sami Area of Finnish Lapland," Acta Universitatis Ouluensis, Ser. A, No. 40 (1976); Geographica No. 2. Oulu.

6. Stig Jaatinen and Pentti Alalammi, "The Field-reservation Scheme of Finland 1969-19777," Nordia 1978, Vol. 12, No. 1. Oulu.

7. Martti Joensuu, "Ruotsiin muutosta ja peltojen paketoinnista Keski-Pohjanmaalla" (Department of Geography, University of Finland, unpublished manuscript (1973.

8. Esko Kalevi Juntunen, "Ruukin kunnan maatalous III. Pellonvaraustoiminta ja maaseudun autioituminen Ruukin kunnassa vuonna 1975," Department of Geography, University of Finland, unpublished manuscript (1976).

9. Tauno T. Maki, "Kemi OY:n yrityskuvan tutkimussuunnitelma," Nordia tiedonantoja, No. 6 (1975): 30-44.

10. Reijo Hakala, "Keminmaan kylatutkimus I, Liedakkala", Nordia tiedonantoja, No. 1 (1979).

9 Agro-industrial Cooperation and the Rural Settlement: The Case of the USSR

V.B. Belenky

Agrarian-industrial cooperation and combination are the most promising forms of the organization of social production in many rural areas of the Soviet Union. In the foreseeable future they may reach the highest level of development in the black-earth zone of the Russian Federation, in the Ukraine, Byelorussia, Moldavia, Transcaucasia, and Central Asia. The organic fusion of farming and industrial production or agrarian-industrial integration is a complex, many-sided socioeconomic phenomenon affecting various aspects of social life. Linked with it is the solution of major socioeconomic problems, such as the elimination of differences between agricultural and industrial labor and the obliteration of distinctions between town and country.

In the Soviet Union the questions of agrarian-industrial integration are increasingly passing from the field of theory into that of broad practical activity. At the same time, the study of its impact on the development of related phenomena and processes is becoming more and more urgent. This applies also to rural settlement. Already today, if we want to frame a correct urban development policy, it is essential to know the effects of agrarian-industrial integration on rural outmigration and commuting, demographic changes and the pattern of rural settlement in the concrete conditions existing in various regions of the country.

Before dealing with population settlement in the conditions of agrarian-industrial integration,

let us look at the organization and distribution of
this type of agricultural production.

Agrarian-industrial complexes (AIC) and large
agricultural complexes of an industrial type
(LACIT), with their higher levels of mechanization,
substantially improve the conditions of work. This
helps to retain the personnel, reduce turnover and
even increase the size of a population. The following concrete examples can illustrate this conclusion.

When a large pig-rearing complex for pork
production appeared at Kalityany (Ukraine), the
rate of population decrease in the central settlement was cut down to one-third.

After a large dairy complex went into operation in the New Idfe collective farm in the
Belgorod district of the Belgorod region, 880
people came to work there, including 500 people in
the 17 to 30 year age group, between 1966 and 1971.
In the same period, only 400 people left the farm.
In a neighboring collective farm, which had no
agricultural complexes, there were no new arrivals
at all, while the number of people leaving the farm
over the same period reached 43, including 41
persons aged between 17 and 30 years.

The same is happening in the Moscow region.
On the Belaya Dacha state farm in the Lyubertsy
district, where a hog-growing complex has been in
operation since 1946, of the 920 people who came to
work there between 1966 and 1970, 690 people
remained on a permanent basis. The population of
the Voronovo state-farm settlement in the Podolsk
district has grown by 644 in the past two or three
years. At the same time, the population of the
Tashirovo state farm, which has the same natural
conditions but no agricultural production on an
industrial basis, grew by 16 between 1966 and 1970.

The influence of industrial-type production on
making the rural demographic situation "healthier"
is quite obvious. What is the explanation? Work
in LACITs does not differ from work at industrial
enterprises. Similar production operations are
performed in either case, cultural standards of
production are high, and the working day is fixed
in both cases. A complex technology of production
requires the use of skilled labor. The central
figure of a LACIT is a specialist with secondary
technical education who operates a system of machines and mechanisms. As a result, on the farms
of the Belgorod region having LACITs, the propor-

tion of unskilled workers declined from 49 to 41 percent, while the proportion of workers with specialized vocational training rose from 42 to 63 percent. The general educational standards of the population have likewise grown. In a period of five to six years, the number of people with higher general and technical secondary education doubled and tripled.

LACITs are, as a rule, diversified production units. Thus, the hog-growing complex of the Kuznetsovo state farm in the Moscow region includes a pig-fattening farm, a combined fodder plant, a slaughter section with refrigerator facilities, and a truck pool. The Voronovo state-farm cattle-fattening complex has as its technological side-lines a combined fodder plant and a regenerated milk factory.

Work in LACITs is becoming more and more attractive, giving greater satisfaction to workers. As a result, the proportion of young workers is rising. In the farms of the Belgorod region, where production has been put on an industrial basis, the proportion of working population in the 18 to 40 year age group has increased by 32 percent within five years. The share of the population permanently employed in agricultural production has grown noticeably. For every 100 able-bodied farm members on farms with a complex and those without it, respectively, 60 to 70 and 35 to 40 are employed in basic production, 20 to 25 and 40 to 45 are employed in urban industry and 5 to 15 and 18 to 20 people, respectively, are not engaged in social production.

So far, we have given examples of large agricultural complexes of an industrial type. In agrarian-industrial integration, there is a still more significant modification in working conditions and in the structure of the able-bodied population. In agrarian-industrial complexes, the general structure of employment is more favorable in comparison not only with traditional collective and state farms, but also with large agricultural complexes organized on an industrial basis.

A conspicuous feature of the sectoral structure of employment in AICs is the presence of a specific group of people employed seasonally in agricultural jobs while working in industry at other times. This is one of the advantages of agrarian-industrial cooperation which helps to reduce seasonal variations in agricultural employment and ensures uniform utilization of labor throughout the year. The age structure of the population reveals a marked growth of the able-bodied group. Changed, too, is the family status of people, with a higher proportion of people with small families of two to four and single men and women.

Agrarian-industrial complexes are character- ized by a considerable increase in the proportion of able-bodied people engaged in the services.

Thus, there is a "chain" of interdependencies: fundamental qualitative changes in the sphere of social production transform social conditions in the countryside, and these, in turn, lead to changes in the composition of the rural population, affecting demographic processes. The rural dweller himself is changing. He is becoming more educated and intellectually developed, with more advanced concepts about the content and way of life and with a higher level of social requirements. In his evaluation of the conditions of life and work and the standard of services, the rural dweller is more and more oriented toward urban standards.

The investigations show that the population makes a higher evaluation of the social conditions of life in agrarian-industrial complexes than at traditional agricultural enterprises. Work in AICs is less arduous, rural workers are satisfied with it, and a predominant proportion of them have no intention of changing employment. The new condi- tions of work are evaluated especially highly by workers in skilled jobs. About 70 percent of workers in unskilled trades regard their working conditions as normal.

At present, a sizable proportion of the rural population of the country has at its disposal household plots of land, cattle, and poultry; that is, personal subsidiary farms (PSF). The PSF is a source supplying a significant part of society's needs in potatoes, vegetables, milk, and meat. The availability of PSFs, their size, and the people's attitude toward them depend on material security, conditions of supply, and the social background of people. In traditional agricultural enterprises and agrarian-industrial complexes, people unanimously agreed that it is expedient to preserve PSFs. Substantial divergences arise only over their size. In AIC settlements most people are in favor of a minimum size of personal subsidiary holdings (only

a land plot and poultry) and a rather numerous
group do not want to have them.

The attitude toward a subsidiary holding influences the choice of the type of dwellings built
in a rural settlement. Indeed, if a rural dweller
wants to have a large plot of land with cattle and
poultry, he will naturally prefer a detached one-
story house. If, on the other hand, he associates a
dwelling with modern conveniences and comfortable
conditions of life approximating urban conditions,
then, given a definite level of material security,
he would choose an apartment in an urban-type
house.

In the settlements of AICs and large industrial complexes, people give preference to apartments
in two- to five-story houses. In the Moscow region,
for example, from 50 to 100 percent of the population wants to live in such houses, whereas less
than 50 percent prefer two-story houses with all
modern conveniences and with household plots of up
to 0.6 hectare. In other villages, one in four
rural dwellers prefers a one-story detached house,
even without conveniences, but with a full-volume
PSF.

On collective and state farms with the usual
form of production organization, nearly all services satisfy only a part of the people's needs.
The level of services is considerably higher in AIC
settlements. A different character, different conditions of work, normal working process, and a
regulated working day create a totally different
pattern of spare time pursuits. There are more
opportunities for household work, caring for children, child rearing, for raising the level of
education, sports pursuits, and more. A matter of
particular importance is the provision of conditions for young people that enable them to continue
studies. Secondary education, special training in
courses offered by secondary and higher educational
establishments, are now becoming obligatory.

As is known, agrarian-industrial cooperation
and combination contribute to more intensive industrial development in the rural areas. Under the
traditional organization of production, too, industrial enterprises are sited in administrative
(rural) districts. With agrarian-industrial cooperation, however, the range of these enterprises is
broader, their role in the economic development of
rural areas is substantially higher, their capacity

grows, and distribution becomes ever more rational. Some industrial enterprises are associated in complexes and localized in focal points of the country. They form a specific urban-development base giving an impetus to the existing industrial and service centers and stimulating the emergence of new ones. By their functions, inter-settlement centers of such complexes belong to agrarian-industrial population centers. Their size (in terms of population) depends on a number of factors, the most essential of which is specialization of the agrarian-industrial complexes.

It has been estimated that the centers of agrarian-industrial complexes may have populations as follows:

• 8,000 to 10,000 in areas of developed vegetable growing, horticulture and viniculture

• 10,000 to 12,000 in areas of grain production, potato growing and flax cultivation and also in areas specializing in corn and oilseed cropping

• 15,000 to 20,000 in areas of beetroot cultivation

Rural settlement under agrarian-industrial cooperation and combination is strongly influenced by changes in intersectoral technological links. They become stable, intensive, and localized in a limited number of directions. Inter-settlement production links become more developed, forming the basis of local systems of group settlement. They are like a framework connecting the settlements of an agrarian-industrial complex into a single system of group settlement.

Special economic calculations are made to determine the radius of a group-settlement system, the rational capacity of an industrial complex, and the size of a resource zone of an agrarian-industrial complex of a given specialization for a specific area of the country.

Depending on the type of agrarian-industrial complex and, above all, its specialization, the radii of local systems of group settlement will be as follows:

• 25 to 30 kilometers in areas of developed vegetable growing and horticulture

• 35 to 40 kilometers in areas specializing in corn cultivation

• 20 to 30 kilometers in areas of potato cultivation

• 30 to 40 kilometers in flax-growing areas

• 30 to 35 kilometers in areas of oilseed cropping

• 25 to 30 kilometers in beetroot growing areas (in the southern zone of the country) and 50 to 70 kilometers (in the central regions of the Russian Federation)

• 50 to 70 kilometers in areas of grain cultivation and animal husbandry

Apart from technological inter-sectoral links (such as supply of raw materials, shipment of finished products, or processing of waste), an important role is played also by the character of migration of labor employed in agriculture and industry. Today objective preconditions exist which require a more active use of seasonal man-power resources of agriculture and processing in-dustry, contributing to the emergence of a specific group of rural population occupied seasonally in agriculture and industry.

Out of the total working population employed by agricultural jobs, about two-thirds are engaged in seasonal plant farming--for about three nomths out of a year. During the rest of the year, many workers of this group form a seasonal manpower reserve. Naturally, questions concerning their full-time utilization should be raised.

A part of these labor resources may be em-ployed in building and repairing housing and public and production premises. Another part (about 10 to 15 percent) may work at agriculture-based enter-prises; still another part may be used at industri-al enterprises not connected directly with agricul-ture (production at such enterprises should be labor-intensive and require no expensive equip-ment).

The advantages of this solution to the problem of rural labor resources are quite obvious and consist of the following:

1. Labor requirements of all sectors of the national economy are satisfied without impairing agricultural work and without incurring big additional investments.

2. Conditions are created that induce the most skilled workers to stay in rural localities.

3. In connection with the industrial development in rural areas, a part of additional investments into nonproduction construction may be used to ameliorate difficult rural conditions of existence.

4. Conditions are provided for realizing the ultimate goal of agrarian-industrial integration--the synthesis of agricultural and industrial labor.

Such utilization of labor resources in rural areas stimulates commuting--daily trips by seasonal workers to inter-settlement industrial centers of an administrative district. If, at the same time, some of the permanent employees of enterprises located in these centers take up residence in rural settlements, that is, make daily commuter trips, inter-settlement labor connections will expand considerably. In combination with the inter-settlement trips of rural population for cultural and other purposes from the periphery to local centers, they form the structure of local systems of group settlement.

Investigations conducted so far show that in agrarian-industrial complexes there may be two models of structure of local settlement systems:

1. A one-stage model when the local group settlement system is divided into inner-economic subsystems without any intermediate links; and

2. A two-stage model when the local system is broken up into cluster subsystems which, in turn, are divided into inner-economic subsystems.

The first model corresponds to agrarian-industrial complexes with relatively small zones of production gravitation and to areas with favorable natural and climatic conditions and developed transport links--the southern regions of developed vegetable growing, horticulture, and viniculture.

The model may also be applied in some potato-grow-
ing regions in the central part of the Russian
Federation.

When the zone of production links is rather
extensive (for example, in beetroot growing areas)
and the transport links are poorly developed, the
need arises for forming additional subcenters and
subsystems of settlement to provide cultural and
other services. These subcenters are known as
cluster settlements. They have an estimated population of 2,000 to 5,000 and a radius of gravitation of 10 to 15 kilometers. The structure of
group-settlement systems in such areas is shaped
according to the second model.

The exploitation of additional reserves af-
forded for boosting production by agrarian-industrial integration is also important from the standpoint of shaping inner-economic settlement systems
in administrative units. Let us examine the example
of beetroot-growing areas in central regions of the
Russian Federation, where the growth of technical
standards of agricultural production depends to a
considerable extent on the use of additional labor
resources. It is estimated that agrarian-industrial cooperation here may increase labor provision
for agricultural production by 15 to 20 percent,
thanks to the use of seasonal labor resources of
industrial enterprises such as sugar refinery or a
dairy factory. Accordingly, this will make it
possible to raise the concentration of sugarbeet
crops, the specialization and productivity of labor, and the yields of agricultural crops and to
reduce the costs of agricultural output. Besides,
in this case better use is made of the plant of
sugar refineries and of the waste products of sugar
production (such as feed for cattle or fertilizers). But the most striking fact is that agrarian-industrial cooperation creates a rational redistribution of facilities between industry and
agriculture, part of them being used additionally
to consolidate the material and technical base of
agriculture. This is a characteristic development
trend of all AICs.

Thus, owing to the maximum use of available
reserves and rational distribution of facilities in
agrarian-industrial complexes, a higher technical
level of farming is achieved compared to the
traditional organization of production. Its specialization, concentration, and intensification

grow deeper and the conditions of work are drastic-
ally improved. These shifts in production are
mirrored in the settlement of population.

Higher technical standards of farming and its
deeper specialization lead to the concentration of
rural residents in the settlement so inner-economic
systems of settlement. This tendency is characteristic of AICs of most diverse specialization. Thus,
in the beetroot growing areas of the center of the
Russian Federation, the level of the concentration
of population under agrarian-industrial cooperation
is 1.2 to 1.5 times higher than in the traditional
organization of production. The same goes for the
agrarian-industrial complexes in Moldavia where the
rational size of growing settlements is twice as
high as in ordinary collective and state farms.

Such are the main features of rural settlements in agrarian-industrial complexes.

Let us now examine in greater detail a settlement system within the agrarian-industrial complex
of the Kalarash district of the Moldavian Republic.
In 1971 an agrarian-industrial association (AIA)
was formed there, based on the cultivation and
processing of grapes. At present the AIA embraces
twenty-one state farms with factories out of the
twenty-three existing in the district. It covers
more than 90 percent of the rural centers and 95
percent of the district's farmland.

A rational settlement system for the district
was substantiated by building variant models of two
different trends of development of social production: through agrarian-industrial integration and
without it.

At first the general structure of the group
settlement system was framed, identifying cluster
centers and the zones of their gravitation. The
zone of accessibility of the district center Kalarash, for which it performs the functions of a
cluster center of cultural and other services, was
defined. The competing centers of the rest of the
territory in the external belt of the district were
studied according to the following indicators:

• The place in the system of inter-settlement
links

• The level of provision of service establishments in a settlement

•Transport expenditures incurred in performing cultural and other links

•The degree of interaction between various centers

The role of the centers in the system of inter-settlement links was defined through an analysis of the distribution of the production base and a sociological polling of the rural population carried out in the district. The processing of the findings revealed the directions and motives of trips, the frequency of applying to various service establishments, and the durations of trips. This was then used as the basis for defining the most intensive inter-settlement links and the centers involved. According to these criteria, two centers were particularly conspicuous - the Onishkany and Bravichi settlements.

The established level of provision of service establishments in settlements under review was determined with the use of the index method. (1) Prior to that, for the settlements that might serve as cluster centers (central settlements of farms), data were obtained showing the level of provision of each type of service (per 1,000 people). Preference was given to a settlement in which this indicator was higher and one which concentrated a greater range of inter-settlement services (trade; medical, cultural, and communal services; and secondary education).

As well as evaluating inter-settlement "activity" of centers and the extent of saturation with service establishments, we also defined the degree of interaction of settlements with one another and with a district center for every sector of the external belt of the district. (2) The higher the calculated indicator of the degree of interaction, the more significant the role played by a settlement in a cluster subzone. The calculations made produced quantitative characteristics of settlements. In addition, transport outlays connected with trips to cultural and service establishments under several variants of the distribution of centers were also defined.

The analysis and quantitative evaluation of indicators in the subsystems helped to single out four rural settlements with the best indicators: Onishkany, Bravichi, Sipoteny, and Vornicheny,

which became cluster centers. Inner-economic subsystems (state farms with factories) were attached
to each cluster center within the radius of normative accessibility.

After that, the structural organization of the
local system of group settlement in the Kalarash
district took its final shape. The nucleus of the
system if the town of Kalarash, which, in turn, is
in a subordinate position in relation to a center
of a higher rank, Kishinyov. The district settlement system is divided into five cluster subsystems; four are headed by rural cluster centers, and
the fifth took shape near the district center as a
cluster subzone. Each of the four cluster subsystems includes three inner-economic sybsystems, and
the cluster subzone developing near the district
center includes five inner-economic subsystems (the
overall reduction of the number of farms is due to
the adjustment of the size and limits of land use
over a longer term).

In this way, a long-term, two-stage model of
the local settlement system appeared. The district
and cluster systems of settlement are characterized
in Table 9.1.

The average size of the population in inner-
economic subsystems ranges from 3,700 to 5,700. The
structure of the forecast district system of settlement provides the necessary basis for substanti-
ating rational inner-economic subsystems of rural
settlement and for optimizing them. Eight variants
were worked out, four for each hypothesis of the
development of social production (with and without
agrarian-industrial integration). The variants were
construed on the principle of successive reduction
of the number of prospective settlements and growing concentration of population in them (see Table
9.2).

The settlement system was optimized according
to a set of economic indicators showing the implementation of a particular variant. Most of the
expenditure items are connected with capital construction and general improvement. Expenditures
take into account capital investments in housing
and welfare construction, general improvement, engineering structures, and communications. The cost
of the unused housing and cultural-welfare construction fund was defined during the removal of
nonprospective settlements, and the value of the
construction fund in prospective settlements that

Table 9.1 Settlement System in the Kalarash District

Name of Settlement	Number of people per square kilometer	Number of Rural Settlements in Subsystems	Area of Subsystems in square kilometers	Size of Prospective population in subsystems (thousands)
Kalarash	90	16	212	18.8
Bravich	123	5	104	12.8
Onishkany	116	9	116	13.5
Sipotany	128	7	110	14.1
Vornicheny	172	7	100	17.2
District Total	135	44	632	77.4

Table 9.2. Basic Indicators of Settlement Variants

	Variants			
	I	II	III	IV
	All Settlements	One or two Settlements per farm	Only one Settlement for District	One settlement per group of farms
Number of Prospective Settlements	44	37	17	4
Number of Prospective Villages	0	7	27	40
Average size of Prospective Settlement Population	1,760	2,100	4,550	19,350

can be put to use and is preserved for the calculated period was estimated. In defining the latter, two indicators of the amortization of premises and buildings for the end of the calculated period were taken into account.

The system of population distribution in the Kalarash district is based on large settlements (populated centers with more than 1,000 inhabitants comprise 66.6 percent of the total). It is obvious that large villages that are not slated for growth will continue to exist for a long time, and provisions have been made for essential outlays on minimal construction and general improvement in order to maintain normal conditions of life for the population in the transitional period. All these outlays are in the category of lump capital investments.

Annual outlays include maintenance expenditures on the running and servicing of prospective settlements, transport outlays on freight handling, and trips to workplace and communal establishments in inter-settlement service centers. They also include losses due to the transfer of land for the development of prospective settlements. Account was also taken of such important indicators (practically never figured into district planning) as production-economic outlays of state farms under various settlement forms.

In each variant, calculation was made of a summary indicator - the amount of overall expenditures indicative of the economic advantages of some variants over others. Final calculations showed that variant III in table 9.2 was the best for agrarian-industrial cooperation and variant II was the best when there was no cooperation.

Thus, the final results of comparing settlement variants in two patterns of the development of production are different. When AIAs are organized, the centralized inner-economic system of settlement is most expedient. In this case, only one settlement will be developing--the central settlement with a population of 4,500 to 5,000 people. Under ordinary conditions, the predominant type of inner-economic settlement is different: the central settlement of the economy and one or two settlements of production sections. Most economical will be settlements with a population of 2,000 to 2,500, or half of its size in AIA.

The intensification and streamlining of inter-

settlement ties in conditions of agrarian-industrial cooperation bring out more clearly the structure of the group settlement system and increase
the concentration of rural population. The system
of population settlement in a district comes to be
based on larger units. The role of the district
center and inter-settlement ties increases. This is
an indication of a developing system of group
settlement in an administrative district as a
whole.

    The example of the Kalarash district gives
some idea of the methods used in solving the
problem of settlement in a rural district under
various organizational forms of production. It
should be borne in mind, however, that, in view of
the zonal multiformity of the development of social
production, it is necessary in each concrete case
to devise ways of perfecting rural settlement in
conformity with local conditions.

NOTES

1. V.R. Belenky, "Concerning the Methods of a
Complex Evaluation of Town-Building Situations,"
Stroitelstvo i arkhitektura Uzbekistana No. 4
(1973).

2. P. Hagget, Spatial analysis in Economic Geography, (Moscow: Progress Publishers, 1968).

10 Agriculture, Modernization, and the Development of Villages: The Case of Poland

Halina Szulc

This chapter is concerned with various transformations of village layouts in Poland after World War
II, which have been caused, above all, by changes
in ownership. (1) The various types of village
layouts can be reduced to some models of transformations and are discussed as a case study of
villages in Polish West Pomerania.

The investigated area lies in the northwestern
part of Poland and covers the voivodships of
Szczecin, Koszalin, and Slupsk. The selection of
this region for the subject of research is not
accidental. Pomerania is a very interesting example
of an area with an ancient settlement network and,
at the same time, an area where the greatest
changes in property structures have taken place in
Poland after World War II due to the accumulation
of land by state farms. The agrarian structure in
Pomerania differs from that prevalent in the remaining Polish territories. Socialized farming here
is the prevailing form: in 1977, socialized farming occupied in the voivodship of Szczecin 73.9
percent of the total agricultural land; in the
voivodship of Koszalin, 73 percent; and in that of
Slupsk, 68.1 percent; the total of socialized
farming for Poland on a whole was only 31.4
percent. (2)

The area of agricultural land cultivated by
socialized farming increases in Poland from year to
year at the expense of individual farming. In 1950,
socialized farming occupied in Poland 10.4 percent
of the total agricultural land; in 1960, 13.1

percent; and in 1970, 16.1 percent. These indices
refer to the state and collective farms as well.
However, the share of collective farms was slight:
in 1950, 0.8 percent; in 1960, 1.2 percent; and in
1970, 1.3 percent of the agricultural land. The
rate of the increase of agricultural land cultivated by socialized farming in Pomerania is higher
than elsewhere in Poland. (3)

The great changes in land ownership that
occurred in Pomerania after World War II are
reflected in the field layouts. These processes
are not the same in all the villages of Pomerania.
This is an effect of differences in the adaptation
of the village to the new forms of socialized
economy, which are caused by many factors; predominant among these factors are (a) the genetic forms
of the villages and (b) the form of land ownership.
For the sake of spatial planning, which should
suggest methods to rearrange the existing settlement networks and rationally use the arable land
taken over from individual farming, it is of the
highest importance that models of the transformation of villages be constructed.

In Pomerania separate settlement forms occur
in zones, which run from the west to the east,
parallel to the coast, in accordance with the
pattern of physical regions. (4) In the northwestern part of Pomerania (in the areas of the voivodship of Szczecin, and of the northwestern part of
the voivodships of Koszalin and Slupsk) the villages are of feudal origin. They are large, compact
villages with buildings surrounding the rectangular
or oval squares (green villages), and the pattern
of roads is nodal. (5)

On the basis of a metrological analysis carried out on the plans from the turn of the
nineteenth century, it has been found that their
sites and fields were planned regularly. (6) This
regularity can be seen even now, because the
farmsteads in the village site are of the same
width or its multiple, whereas the site is concen-
tric in relation to the village boundaries. The
field layout in these villages originates from the
former open-field pattern. The villages form a
regular settlement network and lie at a distance of
about three to five kilometers from one another.
They are of medium size (from 20 to 100 houses),
their buildings are situated amid the fields, and
the pattern of roads is nodal. The fields, though
they are relics of the feudal period, were measured

anew at the beginning of the nineteenth century. Following the emancipation of peasants, their land was disassociated from the manor land: this process has contributed to a relatively good layout of the village, though the structure of the settlement network is rather old. Since World War II the settlement network has not changed greatly.

The only exceptions are the narrow coast strip between Kolobrzeg and Koszalin and eastwards of Koszalin, and the area of Goleniow Forests. Dispersed settlements and hamlets as well as villages of the linear type; that is, the street-and-row villages, prevail there. The villages of the linear types are slightly younger than the dispersed settlements.

Small villages of the hamlet and dispersed-settlement types can be found in the southeastern part of Pomerania. (7) This is due to physiographic and soil conditions. As in the former region, larger green villages with a nodal pattern of roads originating from some earlier historical periods are scattered there in the form of small islands (in the vicinity of Bytow, between Bobolice and Miastko, and west of Szczecinek.

Together with historically shaped layouts of buildings and fields, another factor which has contributed to the adaptation of the villages to the new forms of socialized economy is the form of land ownership. In Pomerania large-scale farming has played an important role and continues to do so.

In socialist Poland, three periods of transformation of ownership structure in Pomerania can be differentiated. The first, from 1944 to 1947, was characterized by the subdivision of landed estates and larger farms and the transfer of land to individual farmers; at the same time, the first state farms were also established. Subdivisions embraced mostly small estates, up to 400 hectares, situated around towns. Individual farmers were assigned from 7 to 15 hectares, or, in cases of specialized or breeding farms, up to 20 hectares. In the second period, from 1948 to 1956, collective farms were set up. They were not very successful. The third period started in October, 1956, and since then many collective farms been dissolved and their land has been given back to individual farmers or allotted to state farms or the State Land Fund.

In Pomerania, it is associated with an in-

creasing variety of socialization forms, and brings
about a larger area to be farmed by the state.

Each period is associated with big spatial
changes in the village layouts; new measurements of
the fields were also necessary. These changes
amounted to an agrarian revolution of a dimension
unknown since the early nineteenth century.

Changes in the village layouts in Pomerania
can be illustrated by some selected examples. The
most common type of village is that in which, side
by side, there are private farms and the State Land
Fund. The farms, the owners of which draw a
retirement pension and have no successors, are
taken over by the State Land Fund.

In the village of Zelazo, situated 25 kilometers north of Slupsk, private and state farms, as
well as the State Land Fund, are side by side. The
village has been affected by numerous ownership
changes after World War II. In 1945-1946, the
village consisted of ten individual farms and a
state farm. The sizes of the individual farms were
quite similar: two farms were each 9 hectares in
area, five were 10 hectares each, two were 12
hectares each, and one was 14 hectares. In 1950,
the first regulation according to established norms
was carried out. Twenty farms, each 9 hectares in
area, were organized. Moreover, two farms, consisting of 4.36 hectares each, were assigned to crafts-
men, and the state farm continued to exist.

In the second stage of transformation, a
collective farm, called "New Life" (Nowe Zycie) was
established on an area of 88.33 hectares; it had
ten member farms ranging from 7 to 9 hectares per
capita. In the third stage, the collective farm
was dissolved (October 1956), and an attempt was
made to come back to the previous state, though the
area of some holdings was increased by land from
the State Land Fund.

In 1960 the next stage of spatial transformations took place. The land was allocated anew, and
that belonging to the state farm was consolidated.
Out of fifteen farms established then, there were
only ten holdings remaining in 1973. Their average
size was from 9 to 17 hectares, usually in four
separate plots. The village also contained a state
farm with an area of 800 hectares. This state of
ownership has survived with some insignificant
changes until now.

In the village of Marszewo, situated at a
distance of 18 kilometers north of Slawno, there

are no state farms at all. Marszewo is an old village, mentioned as early as the thirteenth century, and it is a typical garden village. Pre-war holdings amounted to from 40 to 50 hectares. After 1945, fourty-four farms were established, each from 8 to 15 hectares. After 1949, a collective farm with fifteen members was organized on an area of 150 hectares. The dissolution of the collective farm has caused a new division of land among fifty-one farmers, each of them owning four to five plots.

The greatest changes in the settlement network have been caused by the organization of state farms. After the war, state farms were usually organized on the area of former landed estates. For example, in the state farm at the village of Stanomino, situated 9 kilometers from Bialogard (which was the seat of a state farm combine), the buildings of the former estate can easily be differentiated from new constructions. (8) A characteristic feature of new constructions in state farms is their arrangement into such sectors as lodgings and service, breeding, and production. For example, in the state farm Tymien new constructions are set in among the former buildings which, though adapted for use for the time being, will be demolished in the future. There are also recreation grounds, parks, and sporting grounds. Its fields are consolidated and form big blocks.

As stated earlier, individual farms are at first taken over by the State Land Fund and subsequently assigned to state farms. Lands belonging to farms in small villages, for example, in the former powiat (district) of Bialogard, which are situated on poor soils, far from towns, and are not endowed with elementary services, are taken over by state farms and the buildings are subsequently dismantled. Thus, further changes in the settlement occur by the state farms absorbing more and more land.

In the post-war period, changes in the transformations of former villages and estates are accompanied by the introduction of new settlement forms. For example, in the 1950s a new concept of a collective village was worked out. These new spatial models revert to former patterns of the medieval villages - for example, to former double-row settlements (Waldhufendorf) or to open-field villages. The majority of those concepts have

never been realized. In Pomerania, with few exceptions, neither new rural settlements nor new state
farms have come into being, but former villages and
estates have been transformed in various ways.

In summary, the following conclusions may be
drawn:

1. The various types of former field patterns
and various forms of land ownership are of unequal
duration. In the last three decades, changes of
field boundaries have been quite frequent, while
the boundaries of the farmsteads have remained
almost unchanged. This is true of both individual
and socialized farming;

2. The most rapid processes of spatial transformation occur in the villages where there are
both individual and socialized farms. Every few
years, regulation of fields has to be carried out
in order to consolidate dispersed plots transferred
by individual farmers to the State Land Fund, to be
subsequently taken over by a state farm. This
process makes it necessary to start new measurements of the village and to consolidate the plots;

3. It should be surmised that in Polish
Western Pomerania, as in other parts of Poland,
individual farming will be continued, especially in
big villages with larger individual farms. In such
areas, individual farms often increase their land.
This is caused by the state policies of strengthening incentives conducive to the development of
production, introducing small-scale mechanization,
and providing agricultural implements for the private farms;

4. Small villages with unprofitable farms,
situated on poor soils are going to disappear. The
population will seek employment in industry and in
the tertiary sector. The abandoned land will be
taken over by state farms. The replacement of
individual farms by state farms in Western Pomerania is more rapid than elsewhere in Poland and will
become even more so in the future.

Transformation of the rural settlements in
Poland are conditioned by genetic forms, the rates
of change, the preserved forms of land ownership,
and models of new villages. Studies of the types of
transformation are interesting from both cognitive
and practical viewpoints. They are aimed at discov-
ering the degree to which the old and new spatial
patterns overlap one another and also at deducing

the extent of possible adaptation of the former layouts of fields to the new forms of mechanized and large-scale agriculture.

NOTES

1. H. Szulc, "Maaseudun asutusmuotojen muutoksista Puolassa," (On changes in the forms of rural settlement in Poland), Terra 89, No. 4 (1977): 205-210.

2. Rocznik Statystyczny GUS, Warszawa 1978, Table 11.

3. J. Kostrowicki and R. Szczesny, "Przemiany struktury przestrzennej rolnictwa w Polsce w latach 1960-1970," (Transformation in the spatial structure of agriculture in Poland in 1960-1970), Biuletyn KPZK PAN,No. 87: 91-128;
W. Tyszkiewicz, "The transformations in the agrarian structure in Poland 1945-1970," in Transformation of Rural Areas: Proceedings of the First Polish Yugoslav Geographical Seminar, Ohrid, 24-29 May, 1975, J. Kostrowicki and W. Tyszkiewicz, eds. (Warsaw: Polish Academy of Sciences' Institute of Geography and Spatial Organization, 1978): 123-140;
A. Szemberg, "Przemiany w ukladzie przestrzennym struktury agrarnej w Polsce," (Changes in the spatial pattern of the agrarian structure in Poland), Studia KPZK PAN," 5, No. 55 (1976);
D. Gospodarowicz, "Zmiany w osadnictwie pod wplywem gospodarki wielkoobszarawej na terenie Pomorza Srodkowego w latach 1950-1974" (Changes in settlement under the influence of large-scale agriculture in Central Pomerania in 1950-1974) in his Dokumentacja Geograficzna (in print).

4. J. Kodracki and J. Ostrowski, "Regiony fizycznogeograficzne," (Physical-geographical regions) The National Atlas of Poland (Warsaw, N.P.), map No. 41, 1:2,000.000.

5. M. Kielczewska-Zaleska, "Typy osadnictwa wiejskiego" (Types of rural settlement), The National Atlas of Poland (Warsaw, N.P.) map No. 60, 1:2,000,000.

6. H. Szulc, "On the origin and development of small green villages in the region of West Pomerania," in I Paesaggi Rurali Europei (Perugia, 1975): 493-506;
H. Szulc, "Regular green villages in Pomerania," Geographia Polonica, 38 (1978): 265-270.

7. F. Uhorczak, Osadnictwo (Settlement) 1: 1,000,000 map, (Warszawa, 1957);
D. Bodzak and M. Kielczewska-Zaleska, "Wiejskie osadnictwo rozproszone" (Dispersion of rural settlement) The National Atlas of Poland (Warsaw, N.P.), map No. 60, 1:2,000,000.

8. H. Szulc, "Wplyw typu wlasnosci na przemiany przestrzenne wsi na Pomorzu Zachodnim w okresie 1945-1975" (Property structure and types of transformation in the layouts of West Pomeranian villages in the period from 1945 to 1975), Przeglad Geograficzny, 1, No. 1 (1978): 87-99; see figure 10.3.

III

Agrotechniques
and Rural Change

11 The Impact of Dairy Industrial Technology Development on Rural Transformation: The Case of Alberta, Canada, 1900-1980

Marvin T. Sundstrom

The factory system was introduced to the Alberta dairy industry in the latter part of the nineteenth century, in response to the problem of seasonal milk surpluses. Since the transfer of Dominion government creameries to provincial jurisdiction in 1906, the factory system became initially more widespread and subsequently more refined. Technology played an important role in the growth of the dairy industry, with its impact being evident in farm milk production, in transportation of milk from farm to factory, and in the processing of milk. Governments also became more directly involved in the regulation of the industry and in the setting of minimum retail prices for fluid milk. Finally, marketing forces exerted influence on the rate and type of expansion. After seventy years of change, the dairy industry was characterized by a high degree of concentration at the processing level and considerable specialization at the farm level. These industry changes also had an impact on the rural economy of Alberta.

The purpose of this chapter is to identify the technological developments which were instrumental in transforming the dairy industry and to trace the connection between these developments and changes within the rural economy. While institutional and economic factors also played an important role, technological developments were essential in bring-

*Special credit must be given to Doug Phillips, Jim Moore, and Doug Thompson for sharing their vast experience on the Alberta dairy industry with the author.

Figure 11.1.   The Canadian Provinces

ing about change in both the industry and the rural economy.

TECHNOLOGICAL DEVELOPMENTS

A number of general technological advances in the dairy industry have been identified and categorized in Table 11.1. These developments should not be seen in isolation, but as interrelated and part of a broad advance in agriculture. All of these innova- tions originated elsewhere in North America or Europe, and were subsequently introduced to Alberta (see figure 11.1).

Adoption of Technology on the Farm

Among the developments that facilitated the expan- sion of the factory system in the cream and butter industry, introduction of the mechanical separation was of considerable importance. Mechanical cream separators were initially introduced at the factory and operated in conjunction with tributary cream- skimming stations. During the first decade of the twentieth century, mechanical separators shifted to the farm. This enabled the farmer to retain his

Table 11.1. Technology and the Alberta Dairy Industry

Technology	Impact	Period Introduced
	Farm	
Farm cream separators	Facilitated cream collection, improved butter quality and financial incentive to farmers	First decade of the 20th century
Milking machine	Increased potential herd size by reducing labor requirements	1911 at Strathmore, widespread in 1950s
Bulk farm storage and cooling*	Increased volume and quality of milk handled at the farm	1956, widespread in 1960s
	Transportation	
Extension of railroads and roads	Extended markets for butter and increased cream and milk supply sheds	First two decades of 20th century (rail); 1920s to 1960s (roads)
Transport of milk by bulk tanker truck*	Facilitated high volume processing by greatly extending supply areas	1956, widespread in 1960s
	Processing	
Factory system for butter and cheese production	Improved quality and increased volume of production. Used seasonal surpluses of butterfat and milk	Last decade of 19th century, widespread in first three decades of 20th century
Pasteurization of milk and cream	Upgraded quality and health standards for milk. Extended keeping quality of butter	1906 in Edmonton, 1909 in Calgary, widespread in small centers in 1950s
Stainless steel equipment	Enabled processing plant to improve quality and productivity and reduce labor inputs	1930s, widespread in 1950s and 1960s.

Source: Compiled from annual reports of the Alberta Dairy Commissioner and the Dairy Branch, 1906-1977.

*These innovations also involved the use of stainless steel equipment.

skim milk for animal feed and to ship the more valuable butterfat. This fit nicely into the mixed-farming operation that was characteristic of that period. According to Selitzer, the result of this innovation was that "both the volume and quality of butter production increased, and so did the financial incentives for the farmers to pursue the business of dairying." (1) Much of the butterfat separated on the farm was utilized in the production of farm dairy butter that supplied local markets.

The introduction of milking machines on Alberta dairy farms was first reported on by the dairy commissioner in his annual 1911 report:

> The milking machine in use at the demonstration farm of the Canadian Pacific Railway at Strathmore, is one perfected by the D.H. Burrell Company of Little Falls, New York. It is run on the vacuum system and appears to give as near perfect results as possible. We used this milking machine for three years in Montana and are well satisfied with it. In fact, we are thoroughly convinced that the machine is the thing that is going to solve the great labour and drudgery problem of the dairy business. (2)

In spite of their obvious advantages, milking machines did not become widespread in the Alberta dairy industry for several decades. Resistance to change, difficult economic times, and the smaller mixed-farming operations made for a slow spread of milking machines.

The largest number of milk machines on Alberta dairy farms was achieved in 1966. Since then the decrease in milking machine numbers has reflected high attrition rates among dairy producers (table 11.2). In this process, many small mixed farms were replaced by larger more specialized fluid milk farms. Many farmers, who initially invested in milking machines, found this was just one step in the continual process of specialization and structural adjustment. In the end, they opted out of the dairy industry.

Another important farm innovation occurred in the 1950s with the introduction of the bulk milk tank. This provided farm storage and cooling of milk, greatly reduced farm labor requirements, and facilitated the transfer of milk from farm to

Table 11.2. Number of Milking Machines Reported
   for Alberta Dairy Farms 1931-1976.

Census Year	Farms Reporting
1931	366
1941	Unavailable
1951	4,134
1961	9,238
1966	10,165
1971	8,167
1976	5,917

Source: Census of Canada, 1971, Historical
          Series, Statistics Canada.

tanker truck. The introduction and rapid spread of
bulk tanks in rural areas was promoted by dairy
processors who realized quality improvements and
reduced handling costs as a result. Rural electri-
fication had previously become widespread in rural
areas, making the introduction of bulk tanks feasible. The annual report of the dairy branch noted
the development:

> The system of handling milk with farm bulk
> tanks and milk transportation tanks introduced
> in the Edmonton district during 1956 expanded
> considerably in 1957. It is estimated that
> well over 100 farms were using bulk tanks at
> the year end. One distributing plant in
> southern Alberta is now receiving all of its
> milk by tank. This method will undoubtedly
> tend towards greater specialization in the
> production of fluid milk. (3)

Although bulk farm tanks were initially used to
supply fluid milk plants, by 1976 some 1,070, or 55
percent, of the 1,930 farm bulk tanks in the
province were shipping to industrial milk plants.
(4) Consequently, bulk farm tanks were widely
adopted by farmers who were supplying milk for dairy
products such as cheese and ice cream. Adoption of
this technology throughout the industry lead to a
reduction in the quality differences that had formerly existed. This increased pressure for the

elimination of the distinction between milk producers on the basis of the market they were serving.
However, the most important impact of bulk-tank
adoption was to accelerate farm size adjustments and
increase specialization, as the dairy branch had
predicted in 1957.

Transportation and Technology

Improvements and new developments in the field of
transportation extended over the time period covered
in this chapter. While much of this impact was a
result of the extensions of railway lines and roads,
technological changes were also playing a role. The
route of the Calgary-Edmonton railway line had an
initial impact on dairy industry location, as Prevey
noted:

> The government-managed farmer-owned cream-
> eries were located, for the most part, close
> to the Edmonton-Calgary railway line. Innis-
> fail, Red Deer, Olds, Westaskiwin, Blackfalds,
> Markerville, Ferry Bank and Lakeview were main
> towns and villages where these creameries were
> established. This was to be expected as it
> was essential to have good rail connections to
> Calgary where the Government assembled the
> butter in carload lots for shipment to B.C.
> (5)

Railway branch-line expansion opened up new supply
areas and permitted a limited number of large
centralized creameries to extend their supply areas
through aggressive buying of cream. By the 1920s,
roads were becoming increasingly important in making
cream supplies available to local farmer-owned
creamery associations. When cream-buying stations
were prohibited in 1922, local creameries increased
in number and reached a peak of 100 for 1927 and
1928. (6)
    Increased urban demand and transportation re-
strictions brought on by the Second World War were
to lead to a brief revival of rail transportation,
as Patterson noted:

> Prior to the outbreak of World War II, the
> Calgary milkshed had a radius of about 20 miles,
> but 80 percent of the required milk supply came
> from the rolling land north and west of the
> city. Extraordinary wartime demands, however,

necessitated an extension of the area to 50 miles, and milk coming to Calgary from that distance is usually shipped by rail. (7)

Following the war, transportation restrictions were removed and truck transport once again increased in importance. Introduction of bulk-milk tanker trucks along with farm bulk tanks in the 1950s and 1960s had the effect of extending milk sheds for dairy plants and facilitating high-volume processing. It also established highway transportation as the exclusive means of moving milk and cream from farm to factory. These developments placed considerable pressure on small plant operations and lead directly to consolidation of the processing industry in the 1960s (see Table 11.3). In addition, it had the effect of gradually eliminating milk producers who were unwilling to adopt this new technology. The impact was most evident in the 1960s.

Table 11.3. Alberta Dairy Farm* Trends 1951-1976

Census Year	Farms Reporting	Percent of Previous Census
1951	59,617	90.3
1956	52,396	87.9
1961	44,748	85.4
1966	33,234	74.3
1971	21,368	64.3
1976	13,993	65.5

Source: Census of Canada, Agriculture, Various Issues, Dominion Bureau of Statistics; Statistics Canada.

*Includes all farms with milk cows, thus overestimating the number of commercial farms.

Technology in Dairy Processing

The factory system made its appearance in Alberta for cheese and butter production in 1888 and 1890, respectively. Additional creameries and cheese factories were opened in the last decade of the nineteenth century to deal with the problem of seasonal milk surpluses. Fluctuations in the number of patrons, the extensive nature of early agriculture, and management problems made for considerable turnover in these early creameries. By 1897 the Dominion government became involved in the management of creameries in order to provide stability to the industry and to open new markets. Privately operated plants subsequently entered the industry, and by 1918 a limited number had achieved a dominant position in the dairy industry through high-volume processing. Dairy farm butter, however, remained an important part of the industry until the 1940s, when a Dominion government subsidy program for creamery butter production accelerated the decline of farm butter production.

The quality improvements and uniformity in butter production arising from the factory system opened up external markets, especially in British Columbia and the Yukon. (8) More important was the role that high-volume factory production played in the initial expansion of the industry. Several entrepreneurs were quick to realize the importance of high-volume processing in lowering unit costs, and proceeded to build large centralized creameries, drawing on large areas and using railway shipment for the collection of cream. Elimination of cream-buying stations in 1922 brought about decentralization and expansion in smaller population centers. Some of this expansion was due to new plant construction by the large creameries.

Pasteurization of milk and cream made its first appearance in 1906 and 1909 in Edmonton and Calgary, respectively, when W. W. Prevey and the Carlyle brothers began selling pasteurized milk. Although pasteurization of fluid milk was introduced at an early date in major centers, it was not until the late 1940s and early 1950s that the practice became widespread in small towns in Alberta. This resulted in a number of milk departments being established in many of the creameries. However, by the 1960s many of these plants were being closed down, with major city plants supplying the needs of rural communities. The savings in unit production costs resulting from large volume throughout more than offset the additional transportation costs, thereby contributing to this consolidation.

The importance of pasteurizing cream for creamery production became evident about 1910, but was not widely introduced into the factory system until 1916 and 1917. (9) Better keeping qualities and a more uniform product, along with a strong provincial government leadership, produced rapid acceptance of pasteurization in the cream and butter

trade.
Direct benefits from pasteurization of both milk and cream appeared in quality improvement and particularly in much improved health standards. Although the cost of equipment for small plants did not act as a barrier to introduction, once economies of scale became important, only a limited number of firms were able to take advantage of high-volume processing.

While stainless steel equipment was introduced into the Alberta dairy industry in the 1930s, the replacement of wooden churns by stainless steel was still occurring in the late 1950s. The development of stainless steel was to have a major impact throughout the industry.

By the late 1930s the initial problems associated with stainless steel had been resolved, and it was recognized as the product of the future. The advantages of stainless steel are its lack of reaction with acids in milk and its strength and longevity. It helped eliminate aftertaste in dairy products and it made costly retinning unnecessary. More important, however, was the fact that stainless steel was ideal for the clean-in-place technology that was developed in the dairy processing industry in the 1950s. (10) Stainless steel also had application on the farm and in transportation of milk. Consequently, it played an important role in the introduction of farm bulk tanks and milk tanker trucks. Thus, the initial development of a new product spread to all areas in the industry and laid the basis for a number of additional applications.

The above list of technological advances in the dairy industry is by no means complete, but it is intended to give some indication of the types of changes that were occurring within the industry. These changes have all worked toward increasing the scale of operation at both the farm and factory level.

TECHNOLOGICAL ADVANCE AND RURAL CHANGE

The technological innovations in primary production, transportation, and processing of dairy products resulted in higher levels of capitalization and increased productivity throughout the industry. This had a number of implications for rural change.

The establishment of a dairy farm industry provided the basis for a viable cream and butter

industry. The introduction of the farm mechanical
cream separator enabled many small mixed farms to
sell cream on a regular basis to creameries or to
produce dairy farm butter for the local market.
This provided a source of cash income, which was
especially important during the economic depression
of the 1930s. While the cream separator fit well
into the mixed-farm operation, widespread introduction of milking machines and farm bulk-milk tanks
promoted the move toward specialization and increased dairy herd size. When this occurred, small
mixed farms with cream as a major sideline gradually
declined in importance. The rate of decline greatly
accelerated in the 1960s and 1970s, and today
farm-cream shipment has a limited future. Meanwhile,
cream has been increasingly supplied through the
skimming off of excess butterfat from fluid milk,
much of which is sold as 2 percent milk.

Transportation innovations increased the distance of cream and milk transportation and greatly
expanded the size of the cream and milk supply
areas. Initially, railroad lines played an important role in the increased shipment of cream. With
the exception of a period during the Second World
War, truck transportation became increasingly dominant as the road network was improved and extended.
The result was that a smaller number of producers
shipped larger volumes of milk greater distances. As
a consequence, social and economic linkages between
the farm and the local creamery or dairy were
severed.

Technological advances in the processing of
dairy products lead to higher volumes and more
automated methods of production. Local creameries
and cheese plants run by individual proprietors
became less and less common. They were replaced by
newly equipped plants that processed larger volumes
at lower unit costs. Improvements in the transportation systems also made it possible to ship finished products back to small centers at economical
rates. Within large cities, dairy plants relo-
cated from high-rent areas near the city center to
outlying industrial parks.

CONCLUSION

Technology did not operate in isolation in bringing
about the changes discussed above. Rather, it was
only one factor, along with institutional and mar-

keting pressures and economic conditions, which brought about rural transformation.

The technological impact operated more or less continuously over the study period, with the exception of the Second World War period. One can note a lag between the period of first introduction and the time of widespread acceptance. The rural impact that occurred generally lagged after the widespread acceptance of a new technology. This was most clearly illustrated in the introduction of bulk milk transportation.

The process of consolidation that is evident in the dairy industry is not unique to that sector of agriculture, but is representative of the more general trends that have been occurring within agriculture. Transportation developments, such as railroad and road extensions, and the introduction of farm bulk tanks and bulk tanker trucks had a most direct impact. These increased the distance that milk was transported. This, in turn, made specialization at the farm level necessary in order to complement the high-volume processing facilities. In addition, it made the shipment of processed dairy products from large central plants to smaller centers economically attractive.

The major effect of many of the technological innovations was to increase the level of capitalization at both the farm and factory. This raised the barriers to entry in the industry, and over time it tended to eliminate the producers, transporters, and processors who were not investing in productivity-increasing technology. Consolidation and centralization occurred in the processing industry, while attrition and specialization emerged at the farm level.

The process of economic rationalization that led to the consolidation of both the farm and processing sector of the industry has not been without its detrimental effects. As centralization occurred, the dairy processing industry moved out of small centers, resulting in losses of employment and municipal tax revenue. Linkages between the producers and their local dairy creamery were broken. Increasingly, dairy products were supplied to the towns by large city plants. The number of milk producers decreased sharply as specialized high-volume, technologically advanced, and capital-intensive dairy farms complemented the milk transportation and processing industry, which had already made that advance.

NOTES

1. Ralph Selitzer, The Dairy Industry in America (New York: Books for Industry, 1976), p. 82.

2. Annual Report of the Department of Agriculture of the Province of Alberta 1911, Report of the Dairy Commissioner (Edmonton, Alberta, 1912), p. 121.

3. Annual Report of the Department of Agriculture 1957, pp. 3-4.

4. Annual Report of the Department of Agriculture 1976, p. 5.

5. Chester F. Prevey, "The Development of the Dairy Industry in Alberta" (unpublished M.A. thesis, University of Toronto, 1950), p. 52.

6. Annual Report of the Department of Agriculture 1928, p. 114.

7. H.L. Patterson, Dairy Farm Business in Alberta 1939 to 1943 (Ottawa Government Publication No. 812, 1948), p. 13.

8. Annual Report of the Department of Agriculture 1909, p. 71.

9. Annual Report of the Department of Agriculture 1916, p. 73.

10. Selitzer, The Dairy Industry in America, pp. 365-368.

REFERENCES

"Alberta Dairy Processing Firms," The Canadian Dairy and Ice Cream Journal (August, 1955): 54, 55-60, and 66.

Alberta Dairymen's Association, Historical Notes on Alberta Dairying to Commemorate the 50th Annual Meeting of the Alberta Dairymen Association (Edmonton, Alberta, 1969).

Annual Reports of the Department of Agriculture of the Province of Alberta 1906 to 1977, Report of the Dairy Commissioner 1906 to 1946 and Report of the Dairy Branch 1947 to 1977 (Edmonton, Alberta).

The Calgary Herald, February 6, 1939. Special Section on the Alberta Dairy Industry.

Census of Canada, Agriculture, various issues. Ottawa: Dominion Bureau of Statistics and Statistics Canada.

Craig, G.H., J. Proskie, and V.A. Wood. "The Production of Fluid Milk in the Edmonton and Calgary Milk Sheds, "Scientific Agriculture, 17, 7 (March, 1937: 401-419.

Innis, H.A., ed., The Dairy Industry in Canada (Toronto: The Ryerson Press, 1937).

McCormick, Veronica. A Hundred Years in the Dairy Industry (Ottawa: Dominion Loose Leaf, 1968).

Patterson, H.L. Dairy Farm Business in Alberta 1939 1939 to 1943 (Ottawa: Department of Agriculture, Publication No. 812, 1948).

Prevey, Chester F. "The Development of the Dairy Industry in Alberta" (unpublished M.A. thesis, University of Toronto, 1950).

Selitzer, Ralph, The Dairy Industry in America (New York: Books for Industry, 1976).

12 The Impact of Fruit Production on Rural Transformation: The Case of Central Japan

Hiromichi Nagashima

The national economy of Japan developed conspicuously since the 1960s. The rapid economic growth was achieved by marked progress in the heavy and chemical industries in particular. The industrialization was concentrated mainly on the Pacific Belt zone stretching over industrial areas from Tokyo to Kitakyushu.

During this period, there has been a large-scale migration of labor from agriculture to other sectors due to the increasing job opportunities in nonagricultural industries. The share of the population working in agriculture has dropped from 26.8 percent in 1960 to 13 percent in 1975. In other words, there has been a sharp depopulation of remote rural areas, in contrast with the problems of overcongestion in the urbanized regions. In about 30 percent of the villages and towns, the population has been decreasing at the rate of more than 2 percent a year, and two-thirds of these villages and towns are located in mountain areas. Because of the high percentage of aged people, the society often lacks a population sufficient to maintain the minimum number of economic and social activities necessary for the maintenance of human habitation.

Agricultural modernization has been promoted by the national government in accordance with the principles laid down in the Agricultural Basic Law of 1961. This agricultural policy is divided into two periods. The first period started in 1961; its main purpose was to enlarge farm size and to reduce the disparity of income between agriculture and

153

other sectors. The fruit, vegetable, and livestock industries were selected as profitable sectors of agriculture. Of course, rice cultivation has been the most important sector for its productivity and profitability.

The second period began in 1970, with the promotion of a comprehensive agricultural policy. The purpose of this policy was to create as many large-scale and highly efficient viable farms as possible, (1) and to encourage the development of production organizations including part-time farmers so that the scope of the unit of operation would be enlarged. (2)

The subject of this book is the impact of modern agriculture on rural transformation. The aim of this chapter is to show the development of viticulture and the organization of agricultural production in the case of Makioka-cho, Yamanashi Prefecture. Makioka-cho is selected as a typical rural town to investigate such a subject.

SITE AND FEATURES OF MAKIOKA-CHO

Makioka-cho is situated in the northern part of the Kofu basin. The area of this town is 10,233 square kilometers, where mountainous woodland covers 82.2 percent of the total area and arable land is limited only to 7.7 percent. The population fell from 9,892 in 1960 to 7,502 in 1979.

Yamanashi Prefecture is famous for its viticulture. The vineyard area was 5,870 hectares in 1978, or 19.6 percent of the total vineyard area of Japan. The output of grapes amounts to 25.8 percent of Japan's total products. As the viticulture of Makioka-cho has developed recently, its features are shown in the high rate of newly planted area.

INTRODUCTION OF VITICULTURE

The structure of agricultural production has changed considerably in this town since the 1960s. This change is shown clearly in the index of agricultural gross income. (3) Agricultural gross income rose from 511 million yen in 1961 to 2,221 million yen in 1977. During this period, the contribution of rice, wheat, and other originally upland field crops dropped while the share of the output of fruit rose to 46.7 percent in 1977 as compared with 7.6 percent in 1961. Fruit and sericulture have become the most important items of this town (see Table 12.1).

The developmental factors of grape production are as follows:

* The increase of demand for fruit as a result of economic growth

* The high profitability of viticulture

* The decrease of profit on sericulture

* The existence of fields converted from paddy fields to vineyards in accordance with the policy of the national government

Japan's agriculture has been faced with a serious surplus problem with regard to rice since the 1960s. A rice production control program was introduced in 1970. Paddy fields were transferred to other crops and nonagricultural land use. In Makioka-cho, approximately ten hactares of paddy fields were transferred to vineyards.

As a result of the development of viticulture, an axis of agricultural production was established. The structure of agriculture in Makioka-cho was transferred from only sericulture to fruit and sericulture. The viticulture, however, was introduced mainly by the farm households of the eastern part of this town (see Table 12.2).

The agriculture of the eastern part depended upon sericulture and paddy field, some of which were owned by landowners before World War II. On the other hand, the income of a farm household in the western part consisted of sericulture and forest management. The average income per farm household was higher than that of the eastern part. At least until the fuel revolution, forest management had a sufficient economic function in this area.

As a result of the introduction of viticulture, the previous relative positions of strength of the east and the west have been reversed. Farm households where the viticulture was developed have begun to carry out the land-improvement projects

Table 12.1. Agricultural Gross Income of Makioka-cho

Item	1961		1975	
	Million Yen	Percent	Million Yen	Percent
Agricultural Gross Income	511	100.0	2,220	100.0
Crops	242	47.4	1,455	65.5
Rice	53	10.4	60	2.7
Wheat, Barley, and Miscellaneous crops	44	8.6	8	0.4
Vegetables	48	9.4	101	4.5
Fruits	39	7.6	1,037	46.7
Crops for Industrial Use	54	10.6	122	5.5
Others	4	0.8	127	5.7
Sericulture	228	44.6	659	29.7
Livestock and Its Products	22	4.3	106	4.8
Others	19	3.7	–	

Source: Statistics of Agricultural Gross Income, (Tokyo, Japan: Ministry of Agriculture, Forestry, and Fisheries, 1976).

Table 12.2. Land Use of Makioka-cho

Item		Makioka-cho	Suwa (East)	Nakamaki (East)	Nishiho (West)
Arable Land (ha)	1960	863.0	315.0	322.0	226.0
	1975	717.0	254.0	274.0	189.0
Orchards Area (ha)	1960	32.4	17.2	14.9	0.2
	1975	144.0	88.0	52.0	4.0
Vineyards Area (ha)	1960	19.0	12.0	7.0	0.0
	1975	86.0	47.0	38.0	1.0
Garden of Mulberry (ha)	1960	357.2	113.9	119.3	123.8
	1975	359.0	88.0	118.0	153.0

for upland fields. (4) The construction of irrigation and roads was emphasized in order to promote
grape growing. The project, covering 187 hectares,
started in 1974. It is estimated that this agricultural investment will enlarge the disparity of land
productivity between the two regions.

AGRICULTURAL MANAGEMENT

The average farm size was 0.54 hectares in 1975 as
compared with 0.55 hectares in 1960, which indicates that there was no significant improvement in
farm structure over the fifteen-year period. More
significant changes in the farm structure have
occurred with regard to part-time and full-time
farm households, with the former showing a steep
increase. The percentage of part-time farm households in total farm households rose from 51.2
percent in 1960 to 73.5 percent in 1975.

As a result of the outflow of labor from
agriculture, the shortage of labor became the most
serious problem. Most of the pupils who graduated
from high school have left their native towns to
enter universities or to work in Tokyo, Kofu, or
other cities in Yamanashi Prefecture.

FORMATION OF GROUP FARMING

Group farming, or production organization, was
considered as a measure to cope with the labor
shortage. Group farming is generally thought to be
a measure for expanding the scale of operations,
but does not imply any change in the ownership of
agricultural land. This change is also needed to
cope with labor shortages in remote rural areas
like Makioka-cho. In such a region, agricultural
production is closely connected with the maintenance of the social life. The most influential
factor to succeed in group farming is the existence
of viable farms. Some viable farms have been
established in Makioka-cho, but organization of
farm households has not been developed. The reasons
are thought to be as follows:

1. The main labor force of a viable farm consists
    of middle-aged or aged farmers.

2. Land improvement is greatly hampered by unfavorable factors, including the slope of the land and highly fragmentary ownerships. Most of the farm roads have been improved during these twenty years, but are still very narrow. The mechanization of agriculture is difficult under such land conditions.

3. It is true that group farming is popular in rice production, which is the most mechanized sector of Japan's agriculture, but the total area of paddy fields is very small in this town. Agricultural machinery is not efficient. Consequently, one of the key factors for production organization does not exist in Makioka-cho.

In addition, even in the case of an operation of rice production that is organized, another sector, such as fruit, vegetable, or livestock industry, is managed individually. Hence, the laborers who were to have been utilized by group farming are concentrated instead on their own small, privately worked lands.

Selective expansion of agricultural production was promoted by the national government during the period of rapid growth of the nation's economy. However, intensive agriculture, or specialized agriculture with large-scale farmland, was not developed sufficiently. On the contrary, agriculture intensified its dependence on part-time farming in order to catch up with the growing incomes in other sectors, instead of improving farm size. Since part-time farms are responsible for a fairly large part of agricultural production, the formation of groups inevitably involves these farms, with full-time viable farms as a core.

The necessity of group farming is easily understood. A new policy relating to regional agriculture has been promoted by the national government, but it is very difficult to carry out, especially in the remote rural areas such as Makioka-cho.

NOTES

1. Viable farms are defined as those farm households whose incomes from agriculture amount to at

least the average income of nonagricultural households in towns and villages.

2. Part-time farm households are defined as those
which have more than one person engaged in off-farm
work from among their family members. They are
classified into two categories according to the
degrees of concentration on farming: Class I includes those earning more than half of their income
from farming; Class II farms earn more than half of
the income from non-farm jobs.

3. The agricultural gross income (total value of
agricultural receipts) is obtained by adding the
following: cash receipts from products sold; value
of products consumed within the household; increased value of livestock and plants; and the
increase or decrease of stock of agricultural
products.

4. The expenses of such a project are shared by
the national government, prefectural government,
and beneficiaries.

REFERENCES

Nagashima Hiromichi. "Influence of the Growth of
    Beef Cattle Raising on Agriculture in Moun-
    tainous Regions: The Case of Nagiso-cho,
    Nagano Prefecture," Kyoyo Ronshu of Kokushi-
    kan University 5, 1 (1979).

Agricultural Policy in Japan (Paris: OECD, 1974).

13 The Impact of Introducing Specialized Crops and Changes in Land Use Patterns: The Case of Israel

David Grossman
Yitshak Ofer

The problem of adapting the early-nineteenth-century model offered by von Thunen (1) to modern farming has aroused renewed concern and interest in the past few decades. The long list of publications which have dealt with the subject testifies to this new interest. (2)

This chapter is concerned only with the purely rural land use spatial pattern. It deals, more specifically, with the patterns evolving in the farmland surrounding the Israeli small-holder village (moshav), whose total area, which is mostly irregular, covers between 3 and 4 square kilometers. Many of the moshavim (plural of moshav) have lately adopted very specialized forms of agriculture, particularly flower-growing in greenhouses; as a result, the former patterns, consisting of either fieldcrops or, more often, uniform citrus groves, have had to be modified.

The hypothesis explored in this chapter is that because of several forces operating in a

*The authors would like to express their gratitude and appreciation to the Bar-Ilan University Research Authority, to Mr. Nahum Freeman and the Ministry of Agriculture, Raanana Sub-District, and to the Extension Service Officers of the Jerusalem Division, to Mr. Asher Kula and Miss Tamar Dror who assisted with the field work and the processing of the data, and especially to the numerous farmers and moshav-secretaries who assisted in the preparation of this research.

farming system, land uses involving high labor inputs tend to be pushed to the outer rings, despite the powerful pull to the inner areas, which is associated with the high costs of the friction of distance.

A modified form of Sinclair's model (3) concerning the reversal of the rings around expanding urban areas may exist, thus, around rural settlements as well. This is partly because of the same "forces" directly mentioned by Sinclair (decline of costs of movements, infrastructure and construction considerations resulting from economies of scale, and the introduction of new techniques), but also because of other factors, which may be only indirectly related to Sinclair's arguments regarding the effects of residential sprawl. In rural areas the equivalent forces are mainly revealed in the preference of spatial cost savings related to the social and economic milieu of the settlement over those associated with farm operation.

Sinclair found it necessary to depart from the "rent" model and introduced, instead, the concept of "value for farming" (VF). In this chapter it is suggested that neither "rent" nor "VF" can be used effectively to explain the spatial structure of the rural areas.

A number of studies, notably those of Richardson (4) and de Lisle (5), suggest that there is a tendency for overuse of inputs in the near fields, and as a result, "economic rent" is lower there than in the outer fields. The methodological approach of Blaikie (6) (which may be viewed as a restatement of the old "intensity theory") that prefers the "principle of minimization of movement rather than the normative concept of rent maximization" seems most useful for pursuing the purpose of this chapter. We will attempt to show that movement minimization is of lesser significance than the forces of residential consideration, scale economies, or of other spatial effects of the modern system.

The fact that labor inputs tend to decrease away from the residential area of the village has been shown to prevail in numerous parts of the world, (7) and it is not overlooked here. This principle was taken into account when the pattern of the settlements, planned for absorbing Israel's mass migrations of the 1950s, was laid out. Each household was provided with a strip of usually 2 hectares extending back from the settler's home

(the "in" plot). In addition, the farmer was
provided with another plot detached from his residence which was assigned to grainfarming or to
other less extensive farming (the "out" plot). In
some cases, the "out" plots were farmed partly or
wholly in common.

The land-use patterns evolving around these
forms of settlements and some of their later, more
compact modifications are dealt with in greater
detail in a separate study. (8) The main finding
of the study was that a "four-ring" model of
alternating high and low intensities tends to
develop where better organization and economies of
scale raise the attractiveness of the outer (third-
ring) plots. This is most likely to happen where
the outer plots are more compact and can be farmed
in large blocks and provided with communal services
for operations such as plowing or harvesting (cases
where the product is also communally owned are not
considered).

Inner plots, particularly in the second ring
(the middle portion of the home strip) suffer from
disadvantages associated with interference of
neighbors' operations (9) or residentially related
land uses, and this results in a "dip" of labor use
in the second ring. The dip may move, however,
even to the first ring, where the size of the inner
plot is 0.5 to 0.8 hectares. In such cases, there
is a tendency to neglect the inner plot except
where very labor-demanding uses, such as flower
growing in greenhouses, are introduced. (10)

Preference for locating greenhouses in the
innermost ring is influenced by the need to utilize
the labor of the wife, whose sensitivity to the
friction of space is higher than that of her
husband. The savings involved in locating at a
place which allows the housewife to participate in
the labor-intensive greenhouse work are difficult
to calculate, but the computations should not be
based solely on the costs of her mobility; it
should take into account, rather, the alternative
cost of household help. Thus, even if the distance
is only as much as two hundred meters, it might be
prohibitive for combining her farm work with normal
household chores. Extra help might thus be needed
in order to release her from her home duties.

Despite the recognized significance of the
spatial factors, the introduction of greenhouses
(usually with plastic rather than glass covers) has

not always raised the level of intensity in the
inner plots. The problems of operating them within
walking distance from the residential area tend to
increase in the more compact moshavim, especially
in those which were founded before 1948 and which
were not subjected to rigorous planning. Even in
some of the more recent ones (1960s), where inner
plots of 0.5 to 0.8 hectares were laid out, the
conflicts between "residentially related uses" and
greenhouse farming have clearly emerged. The following discussion will be devoted to an analysis of
these problems, and will begin with a discussion of
Ram-on, whose inner plots are 0.6 to 0.7 hectares
in size and can be classified as being a modified
version of the compact settlement type.

THE CASE OF RAM-ON

The records of land utilization in the moshav
Ram-on, which was settled in the early 1960s by
second-generation Israelis, show that most of the
"in" plots are practically unutilized, while the
outer ones are cultivated very intensively. Of the
six moshav members interviewed, only one had his
"in" plot cultivated all the time (the crop on this
plot, olive trees, has medium intensity). Three
others had it farmed only in winter, while two
others had it empty or in woods for the whole year.
     The reason for this underutilization was main-
ly the small size of the plot, which was reduced to
some 0.3 hectares. Farm structures, such as chick-
en coops, vacant barns, or storage sheds, occupied
the remainder of the 0.6 hectare. The high investment in modern equipment, which partly found its
expression in the large size of these structures,
accounted for the rapid coverage of the area by
these permanent structures, so that in the competition between plants and animals (including humans)
for the use of this land, the plants lost and the
land was put to uses which did not reflect its
agricultural-productive qualities.
     The case of Ram-on provides a model for a new
approach to the planning of spatial distribution of
flower growing under cover in Israel. After a long
controversy, the moshav adopted a new, experimental
method of assembling all the greenhouses into a
single locality, situated outside the built-up
area. The advantages of this common effort were

still being heatedly debated in the <u>moshav</u>. Its promoters listed the following points in support of this spatial solution:

1.   Savings on infrastructure: roads, power lines (needed for providing light for some flowers) and heat (usually necessary in the Jezreel Valley during some cold winter nights

2.   Economies of scale

3.   Facilitating sorting and marketing

4.   More efficient use of structures, and the prevention of the loss of space by "dead" structures

The last point was strongly emphasized, and is probably the most important one. Many plots rapidly filled up with structures which, in many cases, were no longer in active use. By building greenhouses in a common locality, the structures could pass from one owner to another smoothly, if, for a variety of reasons, the former owner decided to discontinue production. Savings, both in money and space, would, thus, be substantial.

The opponents of the system usually pointed to the spatial problem involved in the movement to the greenhouses and the underutilization of female labor. One of the physical expressions of this difficulty is the fact that the common sorting space, built near the greenhouse complex was used only at a fraction of its capacity, since most of the growers preferred to do their sorting at home. Transporting flowers by car before sorting results in some damage; but it was felt that the distance from the watchful eyes of family members involved even greater damage from the less efficient work of hired labor and the added danger of theft.

The spatial problem is thus a significant one, but the Ram-on model has already been copied by other <u>moshavim</u>. (11) In neighboring Nir Yafe, the dip in the amount of labor in one of the outer plots, observed in a survey conducted there in 1976, was related to the fact that the land was then being prepared for a complex of greenhouses, constructed under the guidance of one of Ram-on's experienced members.

It was interesting to note that even within

the first year of operation, the problem of using
female labor in the greenhouse expressed itself in
Nir Yafe in the form of the establishment of an
"offshoot" home in the new sorting room of the
greenhouse (about one kilometer from the home).
The grower's wife had a number of kitchen utensils
there, and brought an infant with her in a portable
seat. This type of "compromise" would probably be
less acceptable to the people of Ram-on, whose
cultural background does not favor such "compro-
mises" in standards of infant care.

    The relative sensitivity of the wife's labor
to distance (as compared to that of the hired
labor) can be appreciated from the records obtained
in the mountain villages, which will be discussed
in greater detail below.

    The example of Ram-on was carefully studied by
other moshavim, but was not found to be equally
applicable everywhere. The problem of assigning an
area for greenhouse development is, unfortunately,
most difficult to solve in the places which could
have made the best use of it. This dilemma revealed itself most clearly in the case of Kefar
Hess, a moshav located in the Sharon.

THE CASE OF KEFAR HESS

The lands of Kefar Hess were not subject to
rigorous planning as was Ram-on. It was founded in
1933 and, like most of the settlements established
at that time, it was compact and had irregular
territory around it. Its land was divided equally
among the settlers, but the plots were scattered in
order to provide each member with equal access to
various grades of land. A close study revealed,
however, that the plots had undergone a redistribution as the result of exchanges and purchases. In
terms of distance from the household, there had
developed fairly great discrepancies by 1978.

    As long as most of the plots were devoted to
citrus cultivation, no major problems arose. But
when greenhouse farming started to expand, at the
expense of the citrus cultivation, sensitivity to
the problem of plot distribution became quite
clear. (12) Most greenhouses were established
fairly close to the village, but very few in the
residential area itself. In 1978, nineteen out of
twenty-nine were found to be at a distance which

did not exceed 500 meters (in a straight line) from their owners' homes, and within this distance most of the plots devoted to specialty crops other than citrus were also located.

The need for replanning did not result solely from the fact that land for additional greenhouse construction had become scarce. It was partly the result of the increasing size of the greenhouses. The moshav's enterprising farmers were unwilling to satisfy themselves with the standard size of 0.1 or 0.2 hectares, which official planners had considered as a standard unit of production for owners who depended solely on family labor.

The major problems, however, were related to a growing awareness of the incompatibility of high-quality flower production with certain land uses that prevailed in and around the dwellings. Drainage problems developed because of the water that was intercepted by the roofs of the greenhouses. But the most severe problem was the close proximity of the greenhouses to the turkey coops; it was discovered that dust blown from these coops damaged the flowers. The incompatibility of animals and plants, already observed above, thus, became a major problem that could be solved only by careful replanning of the location of the various sectors of the village. This involved a redistribution of the moshav's plots. But since the area was devoted to tree crops (citrus), the implementation of such a solution proved very difficult, even though the official plan recommended it. A compromise which was offered was to "zone" separate greenhouse and poultry-coop areas on land which was found to be suitable after careful ecological and spatial surveys were completed. The compromise was a modified form of the one adopted by Ram-on, whose experience was carefully studied by the Kefar Hess replanning committee. Communal control was not attempted, but be designating a distinct zone, some of the advantages associated with scale economies and greater efficiency of investment in structures could be realized. At the moment of this writing, no recommendation of the replanning committee has yet been carried out, partly because of the crisis in the flower industry that developed in late 1978.

A parallel solution was offered to the problem of poultry sheds. Existing sites were found to be inadequate, not only because of the ecological problems already mentioned and the nuisance they caused to life within the residential area, but also because of the increasing significance of scale economies in poultry and egg production.

The various replanning solutions proposed for the moshav did not offer, in fact, substantial deviations from the existing pattern, despite the awareness of the severe problems outlined above. It was necessary to take into account existing uses whose removal would involve prohibitive costs. However, a plan recently offered for new moshavim, as against older ones, seems to accept the idea that the optimal solution is a "Ram-on type" separation between residential and farmland uses. (13) Because of the compactness of this moshav type, however, no room is allowed in the residential area for farm operations (including animal raising). The inner zone, thus, becomes a ring of low intensity while the expanded second ring becomes the area of the highest intensity, being the preferred "specialized crop zone." Beyond this ring, the gradation of land use appears to follow the Thunen model, but in its broad outline, the three-ring model resembles Sinclair's model more than von Thunen's.

The sensitivity to distance, revealed clearly in the course of the Kefar Hess study, can be evaluated if data are available on the costs of movement between the house and the greenhouse structures. No such data were obtained in Kefar Hess. In the following section, however, the results of a study which made use of such records will be presented.

THE MOUNTAIN STUDY

In an attempt to measure the impact of distance on greenhouse production, a study of fifty-six carnation growers in fifteen Judean Mountain moshavim was conducted. The growers represented practically the total population of carnation growers in these villages (they were selected carefully before permission for greenhouse construction was granted), and personal acquaintance with them allowed close study of the details of their inputs. The local extension service provided additional information on their level of productivity, while their production records could be obtained from the Flower Marketing Board.

The harsh topography of the study area restricted agricultural production to a limited number of sites, and the distances between greenhouses
and homes were, thus, larger than usual. The
spatial setting allowed a more effective appraisal
of the effect of distance. The nature of the
terrain raised the expense of travel, thus, adding
greater weight to the friction of space. The
prevalence of shallow soil, as well as the large
variety of ecologial conditions, did not present a
problem, because the soil and other physical inputs
could be "imported" with relatively little impact
on the profitability of the industry. (The high
returns from the greenhouses can repay the extra
input of soil in several months.) The high capital
amounts invested in this mountain area provide
useful evidence that capital inputs are nearly
indifferent to spatial frictions.

This study provided direct information on
specific inputs and outputs. Total labor inputs
were obtained from normative estimates given by
local extension officers, but hired-labor inputs
and the wife's labor inputs were obtained directly
from interviews with the growers.

Information on travel costs was obtained by
counting all the trips made to the greenhouses by
the owners either for working or for shipping the
produce (no account was taken of workers' other
trips). A distinction was made, on the basis of
personal acquaintance, between farmers who used
cars for transportation (estimated at a cost of 9
Israeli pounds per kilometer) and those who used
tractors (estimated at 10 Israeli pounds per kilometer). No account was taken of travel by foot or
by any other means. The cost of travel was then
calculated by multiplying the above costs per
kilometer by the total distance traveled by each
farmer per year. This value was deducted from the
total income, which could be obtained from records
of the Flower Marketing Board offices, where flower
sales of each farmer were registered.

The effect of distance on labor inputs was,
generally, found to be low, even for the grower's
wife. The correlation coefficient obtained for the
wife's labor input and distance was $R = -.01$, but
despite the low coefficient, it is significant that
it had a negative sign (as expected). Hired-labor
input, on the other hand, had a positive coefficient ($R = .17$) indicating the increasing use of

this form of labor with distance. (Significance levels were not calculated, since we were dealing with the total population and not with a sample.) The curve of total hired-labor use per 100 meters of distance suggests even more clearly the preference of this labor for the more remote greenhouses.

No direct data could be obtained, besides normative labor values, for the labor input of the grower himself (or his children). In order to obtain a realistic estimate of the level of these labor inputs, it was suggested to use export-grade production as substitutes for labor inputs. These data were, in the opinion of the extension workers, the best indicators of the growers' efficiency. They had the additional advantage of being most completely and accurately recorded in local offices. A similar approach to the difficulty of measuring labor inputs can be found in a recent article by Stevens and Lee, (14) who failed to discover a meaningful yield decline in the Lesotho village that they investigated. The statistical analysis of our data also suggested that there was practically no correlation between distance and productivity. The correlation coefficient obtained for the fifty-six growers' flower production and distance (measured in 100-meter intervals) was R = .08. The positive sign points to a mild rise in productivity with distance, rather than vice versa. It is clear, nevertheless, that the growers' labor is more sensitive to the impact of distance than is hired labor, though less so than is the wife's labor.

Total flower production was also obtained (which included the number of flowers sold to the local and foreign markets, as well as those which were discarded). The value of all sales (total income) was then used for calculating the effect of mobility. By deducting from total income value the estimated mobility costs, the "spatial income" values for each distance interval (100 meters) were obtained. The curve of the spatial income, obtained by averaging the values of each distance interval, appeared to be even less sensitive than labor was to the friction of distance. The correlation coefficient obtained between distance and the spatial-income curve (SI) was R = .18. An identical correlation coefficient was obtained for the net income obtained after deducting other, nonspatial costs (capital cost excluded). The curve obtained for these values, which approximate the economic rent, was naturally at a lower level.

The identity of these two coefficients is the expression of the fact that total costs (excluding spatial costs) were almost identical for all farmers, since, as already pointed out, total labor was estimated by using normative figures. Thus, it is possible to use the spatial-income (SI) value as a surrogate for the better-known economic rent.

The fact that the economic-rent curve and the spatial-income curve do not reveal any significant reduction with distance (within the 2.2 kilometer limit) is not too surprising, and is quite in line with the arguments presented in the introduction to this chapter.

The major factor which explains the fluctuations of the economic-rent curve is, indeed, a human one. The major peaks (at distances of 1,000 to 1,400 meters) belong to a village (Zakharea) whose settlers, of Kurdish origin, are known for their high productivity; the secondary peak (at distances of 600 to 700 meters) is the distance where settlers of another moshav, mostly of Hungarian origin, are located. It can be concluded that the study of the mountain villages reveals that the effect of distance on production is marginal at best, and is more than outweighed by human factors. It is clear, thus, that various other forces considered earlier are capable of outweighing whatever benefits could be derived from the proximity of the greenhouse.

CONCLUSIONS

The findings of the mountain study indicate that income levels are not affected in a meaningful way by distance from the farmers' residences. It appears, however, that spatially related inputs, of which family labor is the major form, have some bearing on land-use patterns, though the impact was found to be lower than expected. No data were obtained on the spatial impact of other inputs such as capital and even fertilizers, whose spatial components are normally low, but the detailed studies of the mountain greenhouse operations suggest that the high capital investment was not deterred by the remoteness of these areas.

The fact that the spatial effect, when measured in monetary terms, was marginal, at least within the bounds of a normal farm territory, is significant when the balance between the spatial factors and those associated with economies of scale and other considerations is made. These other considerations primarily include those which are related to household operations and to social or economic considerations.

Pricing these complex uses was not attempted in this chapter. It should be the subject of a separate, future study. Despite the absence of quantitative data, it can be stated that the assumption that the nonspatial considerations, taken together, more than outweigh the marginal advantages of the spatial ones, can be upheld. This tends to be generally in line with the model proposed by Sinclair for urban fringe areas.

Additional studies that will focus on the total costs of movement within the residential areas are necessary before the issue of farm mobility costs can be fully evaluated. The planners' view that the compact settlement is to be preferred over other types was based on a detailed study of the savings that can be obtained on outlays for infrastructure as against the expenses of moving further away into the farmland. (15) But other residential mobility costs involving daily operations (such as to schools and shops) have to be considered as well. Such a study is expected to reveal whether the three-ring model that emerges from the present study offers better total economies than other models (such as the four-ring model, which was found to be associated with more dispersed settlement types). In any case, it must be recognized that the model associated with the agricultural landscape offered here has to be studied against the residential models, from which it cannot be separated before an adequate decision as to the preferred model for planning is selected.

NOTES

(1) P. Hall, ed., Von Thunen's Isolated State (Oxford: Pergamon, 1966). English translation of J.H. Von Thunen, Der Isolierte Staat in Beziehung Auf Landwirtschaft Und Nationalokonomie (Hamburg: Perthes, 1826).

(2) See particularly: E.S. Dunn, The Location of
Agricultural Production (Gainsville: University of
Florida Press, 1967); M. Chisholm, Rural Settlement and Land Use, 2nd ed. (London: Hutchinson,
1968); W.C. Found, A Theoretical Approach to Rural
Land-Use Patterns (London: Arnold, 1971); W.B. Mor-
gan and R.J.C. Munton, Agricultural Geography (London: Methuen, 1971).

(3) R. Sinclair, "Von Thunen and Urban Sprawl,"
Annals, Association of American Geographers 57
(1967): 72-87.

(4) B.C. Richardson, "Distance Regularities in
Guyanese Rice Cultivation," Journal of Developing
Areas 8 (1974): 235-256.

(5) D. de G. de Lisle, "Effects of Distance
Internal to the Farm: A Challenging Subject for
North American Geographers," Professional Geogra-
pher 30 (1978): 278-288.

(6) P.M. Blaikie, "Spatial Organization of Agriculture in Some North Indian Villages," Transactions,
Institute of British Geographers 52 (1971): 1-40.

(7) Chisholm, Rural Settlement and Land Use.

(8) D. Grossman, "Spatial Curve of Labor Inputs in
Plain Moshavim in Israel" (Research Paper, Bar Ilan
University, 1979).

(9) R. Weitz, The Israeli Village in the Technology Era (Tel Aviv: Am Oved, 1967), Hebrew, pp.
125-126. See also R. Weitz, From Peasant to Farmer
(New York: Columbia University Press, 1971),
pp.48-49.

(10) N. Ben-Asuli, "Land Use Belts Around Agricultural Settlements in Israel" (Paper, Bar Ilan
University, 1977, in Hebrew).

(11) The impact of space on the diffusion of
greenhouses in Arab villages has been discussed by
Y. Shnell, "Arab Agriculture in the Sharon," in A.
Shmueli and D. Grossman, eds., The Sharon (forthcoming, in Hebrew).

(12) D. Grossman, "Land Use Model Around the Older Moshav in the Sharon."

(13) J. Maoz, "The Physical Planning of the Moshav in the Specialization Era" (Paper delivered at the Annual Meeting of the Israeli Geographical Association, Bar-Ilan University, December, 1978), pp. 93-96, in Hebrew.

(14) R. Stevens and Y. Lee, "A Spatial Analysis of Agricultural Intensity in a Basotho Village of Southern Africa," Professional Geographer 31 (1979): 177-183.

(15) Maoz, "The Physical Planning of the Moshav," pp. 93-96.

14 The Impact of Modern Viniculture on Rural Transformation: The Case of Hungary

I. Suli-Zakar

The Tokaj-Hegyalja region on the southern and southeastern slopes of the Zemplen and Tokaj mountains has developed into the best-known and most famous of all vinicultural areas in Hungary (figure 14.1). For centuries viniculture and wine making have been the bases of economy here and, at the same time, the main promoting factors in the development of the settlements of the region.

The prolonged crisis of the wine-producing area, which has been going on essentially since the second half of the nineteenth century, is characteristically reflected in the features of the twenty-seven settlements (twenty-four villages and three small towns) in the territory. During the past one hundred years, the population of the region has hardly increased (Table 14.1) and the infrastructural amenities of the region are lagging far behind the national standard. All this has occurred in spite of the fact that for centuries flourishing vine cultivation and wine production promoted the development of a number of settlements (Abaujszanto, Tallya, Mad, Tarcal, Tokaj, Bodrog-keresztur, Tolcsva, Olaszliszka, Sarospatak, and Satoraljaujhely) into market towns, and for a long time the Tokaj-Hegyalja region was regarded as one of the most prosperous and most densely populated areas of Hungary.

The grapevine diseases appearing in the second half of the nineteenth century (mildew, powdery mildew, and especially the phylloxera disaster of 1885-1889) completely destroyed this most famous wine-producing area of Hungary. The so-called

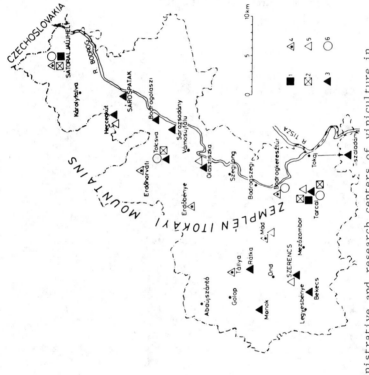

Figure 14.1. Administrative and research centers of viniculture in Tokaj-Hegyalja. Legend: 1. center of the grape producing state farm; 2. district center of the state farm; 3. center of cooperative; 4. center of special cooperative; 5. planned grape- and wine-processing centers; 6. vinicultural and wine-producing research center, laboratory.

Table 14.1. Resident Population and Vinicultural Area
of the Tokaj-Hegyalja Region Between the Years 1869
and 1978.

Year	Resident population of the Tokaj-Hegyalja Region				Vinicultural area of Tokaj-Hegyalja Region	
	Rural		Total			
	Persons	Percent (1869=100%)	Persons	Percent (1869=100%)	Hectares	Percent
1869	44,108	100.00	63,643	100.00	–	–
1873	–	–	–	–	6,025	100.00
1880	39,980	90.64	60,351	94.83	7,769	128.95
1890	42,868	97.19	67,863	106.63	–	–
1894	–	–	–	–	763	12.66
1900	47,816	108.41	79,241	124.51	–	–
1901	–	–	–	–	3,168	52.58
1910	49,808	112.92	86,308	135.61	4,999	82.97
1914	–	–	–	–	5,195	86.22
1920	49,805	112.92	88,190	138,57	–	–
1930	53,085	120.35	90,476	142.16	4,881	81.01
1939	–	–	–	–	4,522	75.05
1941	53,859	122.11	93,463	146.86	–	–
1947	–	–	–	–	3,913	64.95
1949	51,862	117.58	87,579	137.61	–	–
1955	–	–	–	–	4,143	68.76
1960	52,653	119.37	90,728	142.56	–	–
1961	–	–	–	–	4,944	82.06
1965	–	–	–	–	5,749	95.42
1970	49,945	113.23	89,600	140.79	6,267	104.02
1975	–	–	–	–	6,823	113.24
1978	–	–	–	–	7,120	118.17

second reconstruction which began with much hesitation at the turn of the century, very soon ran out of breath. Essentially, only about two-thirds, mostly the lower-lying parts of the former vinicultural areas, were replanted with vine. As a result of the unfavorable internal and international economic conditions, the crisis of the vinicultural region was only exacerbated between the two world wars. The purchasing, storing, and distributing activities that constitute an essential complex within viniculture were taken away from the Tokaj-Hegyalja region. There was a decline in the local background industry related to viniculture and wine making.

After World War II, owing to the profit-withdrawal strategy increasingly applied against Tokaj-Hegyalja, there was a further deterioration in the economic conditions of the region. The fall of interest in wine production and the emigration of the young working-age groups from the territory led, in many places, to the abandonment of vineyards.

The vinestocks planted after the phylloxera disaster grew old by the fifties. In the fifties and sixties, a full reconstruction of the vineyards was needed. After 1955, although with much delay, the replantation of vineyards started. However, due to the changed conditions, the small producers were unable to carry out a newer large-size reconstruction. Among the causes of the failure, a considerable part was played by the unfavorable economic factors: as compared to the vineyards in the Great Plain, the state subsidies to vine replantation were not appropriate, and the purchasing prices were low.

The abrupt shrinkage of the vine-growing area under such conditions could be stopped only by the large-scale cooperative and state enterprises which had the appropriate financial and technical means. However, for these, only the mildly sloping areas, suitable for mechanized cultivation, were utilizable. This is the way in which Tokaj-Hegyalja viniculture "slipped down" to the foot of the hill, to the nearly plain areas along the banks of the river Bodrog.

The third reconstruction, begun at the end of the sixties, has reached its ultimate aim and the Tokaj-Hegyalja region has become more than just a memory of the past. As a result of large-scale mechanized vine-planting carried out on the mild slopes and plain surfaces, the vinicultural region of Tokaj-Hegyalja has regained a territory equal in size to the one it possessed before the phylloxera disaster. Moreover, all previous yields have been surpassed. Presently, the basis of viniculture in Tokaj-Hegyalja is characterized by the grape production of the large agricultural enterprises instead of privately owned, small vineyards (Table 14.2).

As a consequence of social restratification, the proportion of labor employed in agriculture has decreased to 25 to 30 percent in Tokaj-Hegyalja. The number of employees in the local industry (confined primarily to the three localities of urban status) is 14,252, whereas the total number of workers employed in the Hegyalja agricultural large enterprises was 9,327 in 1978. In addition to those employed in local industry, the daily and weekly commuting of industrial and tertiary labor from the Tokaj-Hegyalja settlements to the Miskolc industrial area is very significant.

The number of vineyard owners did not decrease markedly during the restratification, since the agricultural producers who did not move away from the Tokaj-Hegyalja settlements mostly went on cultivating their vineyards. The result of the restratification is manifested rather in the drop of livestock breeding and the simplification of the structure of field crops.

In addition to the small producers, in the area of Tokaj-Hegyalja two state enterprises (Wine-producing Enterprise of the Tokaj-Hegyalja State Farm and the Experimental Station of the National Grapevine and Wine-production Research Institute), twelve cooperatives, and six special cooperatives are engaged in grape production (Table 14.3).

Outstanding among the large agricultural enterprises, owing to its size and importance, is the Wine-producing Enterprise of the Tokaj-Hegyalja State Farm, which was established in 1971. This state enterprise united, on the principles of vertical integration, two enterprises (the State Farm, primarily engaged in vine cultivation, and the Tokaj-Hegyalja Department of the Hungarian State Wine-Cellars Administration, concerned with industrial processing and distribution) into an effectively working economic complex. This enterprise is the first grape- and wine-producing agro-industrial complex of Hungary. The Wine-producing Enterprise plays an integrating role in Tokaj-

Table 14.2. Distribution of Vinicultural Area and Grapevine Production According to the Main Forms of Ownership (1977)

Forms of Ownership	Total Vinicul-tural Area		Total Vinecrops	
	Hectares	Percent	Weight in Centners**	Percent
Agricultural Cooperatives:				
Collective	1,236	17.8	90,300	22.2
Household	1,078	15.5	53,230	13.1
Special Cooperatives*:				
Collective	446	6.4	30,040	7.4
Membership	1,446	20.8	71,660	17.6
Private:	1,215	17.5	40,250	9.9
State Enterprise:	60	0.8	5,200	1.3
Wine-producing State Enterprise:	1,475	21.2	116,000	28.5
Total:	6,957	100.0	406,680	100.0

*Special cooperatives are based mostly on common marketing, but individual cultivation; nevertheless, they have collectively cultivated fields (vineyards), too.

**In the USSR, one centner equals 220.46 pounds.

Agricultural Enterprises	Total Area (Hectares)	Vinicul-tural Area (Hectares)	Average Number of Laborers (Persons)	Total Value of Means of Production	Value of Production (million forint)	Total Ex-penditure	Results According to Balance (1,000 ft.)	Value of Means of Production (1,000)	Value of Production ft.) per Laborer	Results According to Balance (ft.)	Value of Production (ft.) per 100 ft. of Means	Value According to balance
Wine-producing Enterprise of Tokaj-Hegyalja State Farm Satoraljaujhely	3,643	1,465	2,662	1,795	757.5	1,057.1	75,225	674	284	28,258	42	4.1
Agricultural Cooperatives												
1. "Kossuth"- Sarospatak	7,345	109	659	305	168.8	218.7	-1,350	463	256	-2,048	55	-0.4
2. "Hegyalja"- Bekecs	4,308	69	465	128	66.3	71.9	3,613	276	140	7,769	50	2.8
3. "Kossuth"- Monok	3,780	39	376	148	61.5	70.8	2,810	393	163	7,489	41	1.9
4. "Lenin"- Szerencs	3,659	102	362	104	57.9	76.8	-13,808	287	160	-38,143	55	-13.2
5. "Gazdasz"- Olaszliszka	3,348	283	326	136	55.6	56.8	7,510	419	170	23,036	40	5.4
6. "Buzakalasz"- Bodrogolaszi	1,490	50	285	47	53.6	47.5	6,644	165	188	23,312	113	14.0
7. "Tokaj-Hegyalja"- Tarcal	4,398	165	439	106	40.8	54.1	-6,810	243	93	-15,512	38	-6.3
8. "Kemeny"- Hercegkut	2,085	233	239	104	32.4	30.5	5,718	439	135	23,924	30	5.4
9. "Magyar Rona"- Tiszaladany	2,841	49	276	80	29.2	32.4	3,516	291	105	12,739	36	4.3
10. "Buzakalasz"- Ratka	863	90	184	47	20.1	19.2	3,452	260	109	18,760	42	7.2
11. "Bekeharcos"- Tolcsva	569	146	122	44	15.6	16.2	1,592	368	128	13,049	34	3.5
12. "Joremenyseg"- Sarazsadany	910	100	84	32	10.2	13.6	-1,990	382	122	-23,690	31	-6.2
Special Cooperatives												
1. "Rakoczi"- Mad	564	128	145	49	21.2	21.0	4,094	211	146	22,107	42	10.4
2. "Kossuth"- Erdobenye	2,675	72	180	36	20.8	22.7	1,349	202	115	7,494	57	3.6
3. "Beke"- Bodrogkeresztur	1,576	59	132	46	15.9	25.6	2,465	348	120	18,674	34	5.3
4. "Dr. Bartfai"- Tallya	379	64	139	29	6.3	9.3	231	211	45	1,661	21	0.7
5. "Rakoczi"- Erdohorvati	1,839	21	52	12	5.0	5.7	-4	248	96	-76	38	0.0
6. "Egyetertes"- Satoraljaujhely	30	20	36	5	1.3	1.4	0	143	38	0	26	0.0

*Not included among the agricultural enterprises is the Experimental Station of the National Research Institute for Viniculture and Vine-production, although it has vine plantations of 60 hectares.

181

Table 14.4. Grape Production in Tokaj-Hegyalja, 1976-1978, According to the Main Forms of Ownership

Form of ownership	1976				1977				1978			
	Total Vinicult. area (hectares)	Productive vinicult. area (hectares)	Yield of grapevine (metric centners)	Average of crops (metric centners)	Total vinicult. area (hectares)	Productive vinicult. area (hectares)	Yield of grapevine (metric centners)	Average of crops (metric centners)	Total vinicult. area (hectares)	Productive vinicult. area (hectares)	Yield of grapevine (metric centners)	Average of crops (metric centners)
Vineyards of state enterprises	1,513	1,208	104,292	86.3	1,535	1,271	121,200	95.4	1,486	1,324	130,000	98.2
Collective vineyards of the agricultural cooperatives	1,166	1,102	79,312	72.0	1,236	1,159	90,300	77.9	1,337	1,256	92,300	73.5
Collective Vineyards of the special Cooperatives	437	362	30,305	83.7	446	336	30,040	89.4	474	352	25,200	71.6
Small-scale grapevine growers												
Household parcels of Coop members	3,718	3,662	181,685	49.6	3,740	3,684	165,140	44.8	3,823	3,751	192,500	51.3
Vineyards of special coop members												
Privately-owned vineyards												
Total for Tokaj-Hegyalja	6,834	6,334	395,594	62.5	6,957	6,450	406,680	63.1	7,120	6,683	440,000	65.8

Hegyalja, and it is mostly due to its efficient work that viniculture has greatly improved in this region. It helps the new plantations with cuttings and expert advice. With experimental development and expansion of up-to-date agricultural techniques most suited to the geomorphological and pedological conditions, and by lending special machinery, it greatly assists the local vine growers.

The vinicultural area of the district now approaches 8,000 hectares, which can be regarded as the optimal size. However, within this frame, there are sharp differences between the young plantations of the large farms (cultivated with the cordon method) and the old, out-of-date vineyards of the private owners. The big farms produce yields nearly twice as large as the small vineyards cultivated by the families (Table 14.4).

Yields of around 100 metric centners (10,000 kilograms) per hectare are not rare today in Tokaj-Hegyalja. It is also due to large-scale cultivation that the state of the plantations today is better here than the national average.

At the same time, in the opinion of several experts, the quality of Tokaj wines has deteriorated owing to the "narrow" implementation of the third reconstruction. There was no reconstruction in the old vine-growing areas except for resettlement on the mild slopes and plain areas. The fragmentation of the available financial means, in all probability, would not have resulted in such a quick growth of the vine-producing area. However, in the historical wine-producing regions of Hungary, especially in Tokaj-Hegyalja, the quantitative development of production cannot be the exclusive aim. Tokaj wines have a fairly well confined circle of consumption, which demands just this special quality. Thus, the future aim of production cannot be anything other than the improvement in quality, which can only be attained by the replantation of the belts having the most suitable ecological properties for viniculture.

This problem focuses attention on the vineyards of the households and complementary farms. On the old-time Tokaj-Hegyalja vineyard hills, mainly what are referred to as enclosed gardens were formed, and more than 13,000 small-scale growers cultivate an area of 3,823 hectares (in 1978 these constituted 53.69 percent of the total vinicultural region). In respect to Tokaj-Hegyalja, this means that nearly 15 percent of the total

population owns vineyards. Thus, 60 to 65 percent
of families has vineyards even today. In the
villages, naturally, the proportion is even higher: in Mad, Tallya, Tolcsva, Tarcal, and Olaszliszka, 22 to 23 percent of the total population
(80 to 85 percent of the families) possesses
vineyards.

The area of the old-time vine hills can be
divided into three belts according to the size and
state of vineyards: (a) In the lower belt, small
vineyards of medium size can be found; here and
there, one can see vineyards of large enterprises.
The proportion of abandoned vineyards and fallow
land is below 20 percent. (b) In the medium belt,
the cultivated vineyards occupy less than half of
the area. Even these consist of aged vineyards
with lack of stocks. The majority of this belt is
made up of recently abandoned or long abandoned
vineyards, fallow land, and brushwood. (c) The
upper belt was exploited for viniculture only
before the phylloxera disaster. Only a small fragment of the area (10 to 15 percent) was replanted
at the turn of the century, and cultivation practically ceased during the last two decades. Today
this area is covered with an impassable brushwood
of sloe, hawthorn, and briar rose, and one can also
find occasional small patches planted with pine.

On the basis of our representative survey, one
can only get a rather gloomy picture of the
small-size vineyards. Of the privately owned vineyards, for example, only 8 percent are cultivated
with the up-to-date high-cordon method, a little
over 25 percent by the modernized low- and medium-
height cordon; the rest are outdated plantations
giving low yields and cultivated with the levee
method.

Taking the year of plantation into account,
more than half of the privately owned vineyards
would need complete replanting (Table 14.5).

The state of these vineyards is very poor. On
the basis of information obtained from the growers,
the lack of stocks has reached a distressing degree
(Table 14.6).

A considerable number of the interviewed vine
growers (80 to 85 percent) intend to go on cultivating their vineyards. One-tenth would like to
sell them, and the rest (exclusively elderly people) are thinking of abandoning their property.
Only a few, around 1.5 percent of those interviewed, think of replanting their vineyards.

Table 14.5. Distribution of Small-size Vineyards
According to the Date of Plantation

Time of Plantation	Percent
to 1920	30.7
1921–1930	19.2
1931–1950	13.3
1951–1970	29.7
1971–	7.1
Total:	100.0

Table 14.6. Lack of Stocks in Small-scale
Vineyards

Lack of Stocks (Percent)	Vineyard Units (Percent)
0 – 10	24.2
11 – 20	16.7
21 – 30	23.7
31 – 40	18.9
41 – 50	9.1
50 –	7.4
Total:	100.0

Table 14.7. Distribution of Small Vineyards
According to Size in Tokaj-Hegyalja

Size of Vineyards (square fathoms)	Proportion of Vineyard Owners (percent)
0 – 400	11.2
401 – 600	22.1
601 – 800	26.4
801 – 1000	14.4
1001 – 1200	10.7
1201 –	15.2
Total:	100.0

An interesting picture is unfolding from the
distribution of vineyards according to size (Table
14.7). From the data in Table 14.7 it can be seen
that the vineyards can assure appropriate incomes
to only a fragment of the owners, and small-scale
viniculture can give only supplementary occupation
and income. This is proven by the analysis of the
occupations of the heads of vine-cultivating families. Of these, the proportion of those working in
industry is 49 percent; 24 percent work in the
tertiary branches, and 27 percent in agriculture.
This proportion reflects almost completely the
general distribution of occupation.

Any significant amelioration of the state of
small-size vineyards can hardly be imagined. However, the principle of the protection of the
quality of Tokaj wine requires the development of
the places with the best ecological properties,
that is, the utiliztion of just these area.

In the localities that have more favorable
transport facilities (such as Tokaj, Tarcal, Sze-
rencs, and Bodrogkeresztur), recently there has
been an acceleration in the sale of vineyards. The
new purchasers are not the inhabitants of the
Hegyalja region, but mainly urban people from
Miskolc, Nyiregyhaza, Debrecen, Leninvaros, and
Budapest. It can be observed even now that the
change in ownership has resulted in the loss of the
original agricultural function and in the prominence of weekend rest and relaxation.

The government decrees aimed at stimulating
the willingness to cultivate vineyards (such as
freedom from turnover tax on wine for eight years
and free parcels from the national reserve lands)
may have good results, and they could even be
regarded as a significant success if the size and
state of small-size vineyards remained at the
present level. However, some preconditions to this
are still missing. Very naturally, small-scale vine
growers cannot compete with mechanized cultivation
giving high yields. This is why these growers find
the prices of pesticides and agricultural small
machines (necessary for the replacement of manual
work) very high as compared to the purchase prices
of grapes. Moreover, the employment of day labor
for cultivation has almost completely ceased. The
majority of small-scale vine growers avail themselves of the helicopter plant-protection spraying
service (without the helicopter service, the abandonment of vineyards would have been much greater).

However, the growers think it would be important to reduce the prices for this service as well.

The small-size vineyards in Tokaj-Hegyalja mean a constantly decreasing economic source. In spite of this fact, this branch influences the incomes of more than 60 percent of all families. The economic strength of the villages, the wealth of the inhabitants, and the furnishing of the households are significantly enhanced with the additional income from vine cultivation. Any further decrease in the rate of small-scale grapevine cultivation would adversely affect the living conditions of not only those employed in agriculture, but also of those working in industry and in other branches of economy.

The reconstruction of the small vineyards could be promoted by reconstruction loans, by reducing consumption taxes, improvement of the supply of special small machinery, and rearrangement of territorial development, all within the framework of special cooperatives, with replantation in bulk with large-scale methods. It is desirable to give significant support to the special cooperatives, being formed at a number of places, in the form of loans from state funds and through the assistance of large enterprises (wine-producing plants, cooperatives, and special cooperatives, for example), from both the social and the economic points of view).

In recent years, the rate of vine plantation slowed down further, both in the small vineyards and in cooperatives and special cooperatives. This can be explained by the fact that the expenses of production have considerably risen (prices of materials, wages, and ancillary workshops), and this rise was not followed by the rise in state purchase prices (Table 14.8).

In Tokaj-Hegyalja at present, the so-called single-channel order of marketing prevails, which means that the purchase of grapes, must, and wine from cooperatives and small-scale producers as well as the distribution of Tokaj wines is performed exclusively by the Wine-producing Plant. The purchase price of grapes and wine is adjusted by the Plant to the grade of sugar content, which, in view of the high production prices (55,000-60,000 ft/hectare), does not assure the profitability of the cooperatives' vinicultural branches. Thus, a number of grapevine-growing cooperatives are obliged to use up their development funds or even

Table 14.8.   Development of the Purchase Price of
Grapes in Tokaj-Hegyalja between 1971-1976 (Ft/kg).

	1971	1972	1973	1974	1975	1976
From Agricultural Cooperatives	10.62	9.30	10.55	7.29	8.41	6.18
From Special Cooperatives	10.78	10.30	9.48	7.96	9.98	6.30
From Small-scale Producers (household plots of coop members, from members of special cooperatives from private owners	10.93	10.53	11.64	7.39	10.61	7.77
Average Price*	10.80	9.80	10.60	7.40	10.00	6.75

*Notes: 1971:   In the year of its establishment, the Wine-producing Enterprise paid the purchase prices of the State Wine-cellars Administration.

1972:   Excellent vintage.

1974:   Poor vintage. The balance of all cooperatives in Tokaj-Hegyalja was passive in the vinicultural branch.

1976:   The extremely poor quality was partly compensated by the higher yield as compared to previous years.

part of their amortization funds to make up for losses in the vinicultural branch.

The wine-producing activity of the region has been dynamically growing since the establishment of the wine-producing enterprise. In the gradually growing output of the plant, an important role is played by its quickly developing winery. On the other hand, small-scale and cooperative wine making is declining in Tokaj-Hegyalja. Owing to the prevailing law of taxation on wine marketing, the latter are not interested in the storage and maturation of their wines. Because of the drop in the wine production of the cooperatives, a considerable part of the storage facilities remains unutilized. In 1970, wine-cellar capacity was utilized with storage vessels (barrels) up to 50 percent, storage vessels (barrels) were utilized with wine up to 70 percent. In 1975 the respective exploitation of wine-cellars and storage vessels was 45 percent and 20 percent, and the situation has only deteriorated since then. Owing to the decline of wine making, grapevine growing is becoming more and more precarious for the cooperatives. For example, in 1974 and 1976, only those cooperatives that possessed considerable amounts of wine in stock were able to make up for their losses. This is why the cooperatives raised a demand that they should not only participate in vine production, but also have a possibility of participation in wine making and maturing and have an appropriate share in the profit of wine distribution as well. However, to achieve this aim, it is necessary that the cooperatives be given a subsidy (loan) to their working funds according to their wine in stock; otherwise, for the period of the aging of wine (which is three to four years in Tokaj-Hegyalja), they are unable to pay any share to their membership. Without wine in stock, the necessary maintenance and administration of the storage capacity of cooperatives and small producers will be a nonrefundable additional expenditure, which will further worsen their economic results.

In spite of the above problems, the majority of agricultural cooperatives in Tokaj-Hegyalja do not lag behind the local industry as to the efficiency of their economy (Table 14.9) and can assure higher and higher income for their members.

A special place is occupied in the viniculture of Tokaj-Hegyalja by the wine-producing enterprise. Besides grapevine cultivation, due to the above

Table 14.9. Principal data of the Grapevine-Growing Cooperative and the Local Ceramic-Artisans Cooperative in the Village Bodrogkeresztur.

Unit	Average No. of Laborers	Production Value (million Ft.)	Results According to Balance (million Ft.)	Production Value per Laborer (Ft.)	Result According to Balance per Laborer
"Beke" agricultural special cooperative	132	15.9	2.5	120,000	18,674
Ceramic artisans' cooperatives	262	28.1	4.6	107,000	17,557

causes, it performs nearly exclusively the activities of wine making, storage, and distribution.
Since its establishment, the wine-producing enterprise has organized the wholesale and retail marketing of Tokaj wines. Thus, after decades, the
whole vertical organization of grape and wine
production was given back to the region. The
positive socioeconomic effects of this development
are beginning to be felt in the settlements of
Tokaj-Hegyalja.

The 3,643 hectares of the wine-producing enterprise occupy a stretch of 80 kilometers of land
in Tokaj-Hegyalja. It is not at all easy to
organize the work of nearly twenty-seven-hundred
permanent workers (and, in the time of the peak
season, several thousand seasonal workers) in this
vast place of work. The vine-growing area of 1,465
hectares (21 percent of the total vinicultural area
of the region) is divided into four districts. The
status of "district center" gives special importance to such villages as Tolcsva and Tarcal, since
a great deal of the local technical intelligentsia
is concentrated there. In the district centers, in
addition to the administrative and organizational
tasks, actual research bases also have been established where the scientific investigation of grapevine plantation and cultivation is being carried
on. Similarly, the local scientific basis is wid-
ened by the Bodrogkeresztur agrochemical center,
which is being organized now and will be a joint
establishment of the wine-producing enterprise and
the vinicultural cooperatives.

The widespread cultivation and production
technology of the plant, as well as the introduction of the expert advice service, have greatly
promoted grapevine growing both in the cooperatives
and in the small vineyards. By the introduction of
new plantation methods, up-to-date pesticides, and
new ways of cultivation and cutting, the wine-
producing enterprise actually played a pioneering
role in Tokaj-Hegyalja.

In recent years, the wine-producing enterprise
has solved very significant technological, development and research tasks as well. The enterprise is
in possession of 73 power machines, which, together
with the implements, correspond to 150 tractor
units and represent a total performance of 4,000
horsepower. Joined to this stock are 64 trucks
which can satisfy all transportation demands. The
transport of workers has been solved completely in

this scattered, vast, open-air working place, with
their own trucks.

For maintenance, the enterprise runs three
repair shops (Satoraljaujhely, Tolcsva, and Tarcal). The deterioration of machines that have
proven to work out in other vinicultural regions of
the country, is faster here under the special
conditions. On the other hand, some machinery
suitable in plain regions can hardly be used here.

The other producers, owing to the small size
of their vineyards and the smaller number of their
machines, are stimulated to a higher degree of
cooperation. Thus the cooperatives of Olaszliszka
and Bodrogkeresztur established a repair shop
jointly with the wine-producing enterprise. Negoti-
ations are under way between the leaders of the
cooperatives and the plant so as to establish
machinery-repair centers and a joint depository of
spare parts.

In the viniculture of Tokaj-Hegyalja, a great
task is set to the Tokaj-Hegyalja Plant Protection
Association, in which all vinicultural establishments are joined. Responsibility for this task and
the direction of the helicopter plant-protection
service was taken over by the wine-producing enterprise in 1974. Since the spring of 1978, the
spraying of the vineyards of the joint units and of
the contracting small producers is performed by
three helicopters in periods of twelve to fourteen
days, and in times of greater danger of infection,
in periods of eight to ten days. It has been proven
by seven years of experience that the helicopter
has become the basic machine of grapevine protection, and without it the tendency to abandon the
small vineyards located on sites of high slope
angle would be even greater.

In processing wine, the wine-producing enterprise has achieved considerable results. It possesses, at present, four up-to-date grape processing units and a storage capacity of 430,000 hectolitres. The storage and aging of wine is performed
in four wine-cellar districts in nearly two hundred
cellars. This agro-industrial complex also has an
up-to-date bottling plant and a distillery. The
bottling plant at Tolcsva, built at the cost of 80
million forints, has been in operation since the
spring of 1974 and works in two shifts with a
capacity of 100,000 hectolitres per year.

The units listed above and the accessory
workshops (two cooperatives, five wineshops, the

wine depot in Budapest, the depository of pacing
material and bottled wine at Tolcsva, and the
construction unit) do their best to provide the
domestic and foreign consumers in sixty-two countries with an appropriate quantity of wine from
Tokaj-Hegyalja.

The increasing economic results allow the
wine-producing enterprise to spend an ever-growing
amount of money on the professional and cultural
development of its nearly three thousand employees.
It helps the development of the professional knowledge of the laborers with in-service training and
scholarships. The enterprise gives considerable
financial assistance to the cultural homes and
libraries at Satoraljaujhely, Tolcsva, and Tarcal.
It contributes to the building and running of
creches, kindergartens, and schools in several
villages of the region, including Tarcal,
Erdobenye, and Mad. It promotes employee home
building with financial assistance, loans, and
cheap transport.

In the economic improvement of Tokaj-Hegyalja,
the most important stimulating factor is industrialized large-scale agriculture. To a great extent,
it is to the credit of the enterprise, which has
developed into a modern agro-industrial complex,
that economic backwardness in the Tokaj-Hegyalja
region has been eliminated.

The permanently utilizable natural resource in
the economy of Tokaj-Hegyalja is its soil, the more
profitable exploitation and more effective protection of which is one of the most important tasks of
the future. This is supported by the decree of the
Ministry of Agriculture and Food Industry, which
envisages the plantation of an area of 100 to 150
hectares with soil excellent for grapevine growing,
in order to promote the development of the vinicultural region. To promote the recultivation of the
lower levels of the enclosed-garden belt, 70 percent support is granted by the state for replantation and 100 percent for amelioration. (On the
highest level of the enclosed-garden belt, there is
no way of recultivation, due to the very thin, at
places completely missing, surface soil.) For the
protection of soil, a complex plan of amelioration
was developed, including the whole territory of
Tokaj-Hegyalja. Beyond soil conservation, the regulation of waters, afforestation, turfing, resettlement plans, and road building serve the general
protection of the environment, the public roads and

railways, and the interests of the villages lying
at the foot of the hills.

It is worth noting that, whereas Tokaj-Hegyalja has good transportation connections with the
other parts of the country, the internal road
networks of the villages and particularly of the
vinicultural areas are badly neglected. One can
hear more and more frequently that even where there
is a willingness to replant, there is a hindering
factor, namely, the poor accessability of the
agricultural areas. The lack of passable roads is
especially manifest at the times of planting and
vintage. An outstanding role is played, even in
respect to roads of agricultural importance, by the
maintenance activities of the wine-producing enterprise. Nevertheless, some service roads are also
built by the cooperatives as well.

Spraying by helicopter calls attention to
another highly important problem. The helicopters
are supplied with high-quality atomizer heads, and
these require water of pure drinking-water quality
at the twelve places of takeoff. In Tokaj-Hegyalja,
water supply through waterworks is at the beginning
stage; progress in the construction of modern
units, however, is made inevitable by the demands
of modern agriculture.

Further economic measures, such as raising the
state purchase price of must, introducing protective prices, large-scale plantation and replantation, and assisting cooperatives and small-scale
producers by more effective loans, may increase the
willingness to cultivate grapevines. Nonetheless,
the final solution to the economic problems would
be the establishment in Tokaj-Hegyalja, as among
the vine growers, a closed vinicultural system that
would comprise the whole grape-wine vertical complex.

Besides viniculture, the agriculture of Tokaj-
Hegyalja has very significant, and as yet unex-
ploited reserves, such as large-scale livestock
breeding, which could also enhance the dynamic
development of the region.

Presently, the most urgent problems of Tokaj-
Hegyalja are not primarily of an economic character. Just as in other agricultural regions of the
country, the technical development of agriculture
has been much faster since 1960 than has the
development of a rural infrastructure. Thus, between the villages in the Hegyalja region and its
modern agricultural pattern, grave tensions have

developed. The planned and quick release of these tensions is a task of primary importance, but it surpasses the abilities of the local agricultural community.

In the vinicultural units of Tokaj-Hegyalja, economic concentration, mechanization, and the new achievements of biology and chemistry have created favorable technical and scientific conditions for large-scale industrial agriculture. Essential structural and administrative changes have also taken place in this region, in the course of which the labor and marketing conditions of agro-industrial production have been implemented. As a result of the integration of production circles that had previously been separated from one another, the essential administrative, organizational, and industrial servicing and commercial activities have been decentralized and transferred from the rural districts of Hegyalja, to the large-scale agricultural centers.

As a result of up-to-date grapevine-growing agriculture, the demographic structure of the villages in the Tokaj-Hegyalja region has not been distorted. Moreover, the professional level of rural labor has been considerably improved as required by the development of the economy. As a consequence of all of this, the demands of the workers for housing have increased, and they can only be partly satisfied at present by the low-level settlement conditions of the region.

The underdevelopment of rural infrastructure may make the further increase of production questionable in the future.

REFERENCES

Barta, Gy., "Az infrastrukturalis ellatas teruleti elteresei" (Spatial disparities in the provision of infrastructural services), Foldrajzi Ertesito 21, 4 (1972): 459-470.

Beluszky, P., "A lakossag eletkorulmenyeinek teruleti tipusai Magyarorszagon" (Territorial types of the living conditions of Hungary's population), Foldrajzi Ertesito 26, 1 (1977): 87119.

Berenyi, I., "Development of the Agricultural Structure around Kiskoros" in The Changing Face of the Great Hungarian Plain: Studies in Geography in Hungary, No. 9, (Budapest: Akademiai Kiado, 1971): 123-132.

Enyedi, Gy., "A hegy- es dombvideki teruletek foldhasznositasi problemai Magyarorszagon" (Land Utilization Problems in the Mountainous and Hilly Districts of Hungary), Foldrajzi Kozlemenyek 14, 1 (1966): 49-58.

Enyedi, Gy., "A magyar falu atalakulasa" (The Transformation of the Hungarian Village), Foldrajzi Kozlemenyek 23, 2 (1975): 109-124.

Enyedi, Gy., ed., "Rural Transformation in Hungary," Studies in Geography in Hungary, No. 13 (Budapest: Akademiai Kiado, 1976).

Enyedi, Gy., "A falusi eletkorulmenyek teruleti tipusai Magyarorszagon" (Territorial Types of Rural Living Conditions in Hungary), Foldrajzi Ertesito 26, 1 (1977): 67-87.

Enyedi, Gy., "A korszeru falufejlesztes elvei es legujabb tendenciai" (Principles and New Tendencies of the Modern Rural Development), Teruletrendezes 24, 4 (1977): 11-15.

Enyedi, Gy., "A mezogazdasag fejlodesenek hatasa a falusi telepulesekre" (The Influence of the Agricultural Development upon the Rural Settlements), Varosepites 18, 1 (1978): 2-4.

Kapas, P., "A Tokajhegyaljai Allami Gazdasagi Bor-
kombinat megszervezese es tapasztalatai" (Organization and experience of the Wine-producing
Enterprise of the Tokaj-Hegyalja State Farm),
Gazdalkodas 22, 2 (1978): 54-57.

Kulcsar, V., "Falvaink gazdasagi fejlodese" (Economic Development of Villages), Teruleti Statisztika 25, 4 (1974): 353-360.

Romany, P., "A valtozo magyar falu" (The Changing
Hungarian Village), Nemzetkozi Mezogazdasagi
Szemle, No. 6 (1977): 7-11.

Sarfalvi, B., ed., Land Utilization in Eastern Europe: Studies in Geography in Hungary, No. 4
(Budapest: Akademiai Kiado, 1967).

15 The Modernization of Agriculture as a Factor in Rural Transformation: Hungarian and American Analogues

Fred E. Dohrs

The development of modern agriculture has meant almost universally an intensification of farming. Intensification may be defined as increased input of capital and/or labor per unit area of agricultural land. Thus, intensification cannot help but result in substantial changes, and it usually means improvements in farm output, changes in production patterns, and rural transformation.

Modification of settlement forms and patterns and other social changes, which usually accompany intensification of agriculture, often result in the occupational, economic, and social transformation of the remaining rural population. It is clear that increasing output with proportionally higher gross returns cannot help but mean substantially greater income for the smaller rural population.

Over the past ten years, there has been a major trend in the transformation of Hungarian agriculture, resulting from a deliberate decision by the government. (1) This decision meant a strong, even serious, ideological wrench - a conscious turning away from the Marxist-Leninist dogma of industry, especially heavy industry, first, last, and always. It must be recalled, however, that Hungarian agriculture has been in the vanguard of making major modifications of the Soviet model of collectivization, changes that began as early as 1953. (2)

A 1969 decision attempted to maximize production of maize through adoption of the technology and methods of the American corn belt, the most efficient maize-production system in the production system in the world, and apply them to Hungarian

farms. This meant a major shift from what had been
primarily a labor-intensive system to capital-
intensive operations. Furthermore, it meant
large-scale imports of American machinery, equipment, and materials at substantial expenditures of
scarce hard currency. A basic requirement was the
cooperation of American producers, which was readily given, most notably by the high-technology Corn
Production System, Inc., of Illinois. A joint
agency was established in Chicago, which made it
possible to select the most suitable machines for
Hungarian farms, hybrid seed-corn, and the latest
scientific research on farm methods. (3) Necessary
hard-currency credits for essential purchases were
granted by the Hungarian Foreign Trade Bank. Since
1968, state and cooperative farms had been able to
obtain foreign currency to purchase agricultural
machinery and equipment, as well as licenses for
domestic production from the West. Such credits
had been limited and had not been used for a whole
integrated farming system. The credits granted
were to be repaid from increased exports and
reduced imports.

The new program was initiated in 1970 on some
6,000 hectares of the Babolna State Farm in Komarom
Megye (county), and was called the "Industry-Like
Maize-Production System." Even though this was the
first crop raised under a new technology, maize
yields rose to 57.8 quintals per hectare, an
increase of some 23 percent. Total production the
first year was sufficient to pay back the hard
currency outlay for American machinery to the
Foreign Trade Bank.

From this modest beginning, the closed production system (CPS) (4) has grown phenomenally and is
widespread throughout Hungary. (See Table 15-1).

Table 15.1. Growth of the
           Closed Production System (CPS)

	1970	1972	1974	1976	1978	1980
Hectares in CPS (thousands)	6	85	535	1,200	1,450	1,600

Sources: Magyar Hirlap, Jan. 30, 1973, Feb. 1,
    1974; Figyelo, Sep. 3, 1975; Nepszabadsag,
    Dec. 24, 1978.

+Planned

The concept of a high-technology closed system to achieve maximum crop yields by full mechanization of cultivation with the latest techniques, fertilizers, herbicides and pesticides, tested seeds, and reduction of labor inputs is highly capital intensive. During the early years of expansion, the system was broadened to include wheat and sugar beet production as well. On December 4, 1972, the minister of agriculture and food, Imre Dimeny, stated that the closed production system should also include production of potatoes, grapes, and fodder to "achieve an important volume of crop production with marginal natural resources on as many farms as possible." (5)

Under the Babolna State Farm CPS maize operation, the farmed area is divided into a number of production units. Each unit includes about 800 hectares of cropland, together with essential machinery, tractors, plows, harrows, drills, cultivators, and a harvester or combine. The equipment is all matched and integrated and used with appropriate technology with tested hybrid seed, chemical fertilizer, herbicides, and insecticides. A team of six specialists operates each unit. Such a team and field unit, although much larger than optimum-sized farms in the American corn belt, does conform generally with American capital and equipment inputs on a per hectare basis. The realities of rural life on Hungarian cooperative and state farms, however, suggest that additional labor inputs from nonspecialists already on the farms may be quite high. In only a few cases does the CPS include the whole farm. Rather the CPS "unit" has been applied on parts of many farms where there are adequately trained specialists, suitable land, and what would appear to be sufficiently imaginative (even bold) management to make the CPS feasible.

By 1976, a year with poor weather for maize, CPS farms produced 46.6 quintals per hectare, nearly equal to the average yield in the United States. The overall Hungarian average yield was only 38.7 quintals. Increased maize production moved Hungary into fourth place in the world, following the United States, France, and Italy, and 400,000 tons were exported that year. (6)

In January 1979, more than sixty different closed production systems were in operation on Hungarian farms. Of the 141 state farms, 117 operate one or more of the various CPS types, while

1,116 out of a total of 1,425 agricultural cooperatives are involved with at least one CPS. About
one-third of the country's cropland and about 20
percent of the vineyards and orchards are in CPSs.
More important are the facts that the farms operating CPSs have yields well above the national
averages: 55 percent more maize, 36 percent more
potatoes, 29 percent more sugar beets, 125 percent
more soy beans, and 36 percent more sunflower
seeds. (7) These are spectacular results. CPS
production of rice and poultry have proven much
less successful, even unprofitable, and have been
dropped.

Some serious problems hampering the CPS and
further growth of its application result from its
capital-intensive nature. As stated earlier, the
Hungarian Foreign Trade Bank provides hard-currency
credits for foreign purchases. Foreign currency
reserves are, of course, limited, and CPS farms
must compete with other productive enterprises for
allocations of these scarce funds. Through the
"currency credit construction system", farms were
able to buy machinery and supplies directly from
foreign sources. Such direct purchase was abolished
in 1976, apparently because of expressed concerns
of the danger of increased contacts with the West.
The minister of agriculture and food, Pal Romany,
has criticized this view, however, saying that "no
enterprise should fear for Hungary's socialist
integrity just because it has relations with an
enterprise in one of the capitalist states." (8)
Limits on hard-currency purchases nevertheless continue.

Basic to the expansion of the CPS has been an
increased supply of tractors. Neither domestic
production nor production in other Council of
Mutual Economic Assistance (COMECON) countries have
been able to meet the demand. Chronic shortages of
Western currency severely limit outright purchase
of the much-sought-after American tractors. The
Gyor Raba Wagon and Machine Works obtained a
license to manufacture tractors and signed an
agreement with the American International Harvester
firm for joint production of a 245 horsepower
tractor. The first of these tractors reached Hungarian farms in 1975. (9) At present there are
nearly three thousand in service.

Serious problems with COMECON farm machinery
are also a factor in the growth of the CPS in

Hungary. In many cases, COMECON countries are unable to meet the exacting technical requirements of Hungarian users. Deliveries of ordered machinery are frequently below the quantities ordered and are often late. Of a 1978 order for 500 combines from East Germany, only 250 were delivered, and many of these arrived after the harvest. (10) Plus, there have been the almost chronic shortages of spare parts, which idle machinery at the times it is most needed.

Another shortage which seriously constrains CPS productivity is the availability and application of fertilizer. Since the inception of the CPS, application of fertilizer to Hungarian fields has increased dramatically, with 120 kilograms applied per hectare in 1969 and 276 kilograms in 1975 (See Table 15.2). Yet, this by no means meets the need or compares favorably with application rates in more productive farm areas: In 1975, Czechoslovakia applied 305 kilos per hectare; East Germany, 370 kilos; West Germany, 402; and Belgium, 570 kilograms (11).

Table 15.2. Synthetic Fertilizer
Production and Consumption

	1960	1975
Production, 1,000 tons active substance	120	627
Consumption in kilograms per hectare	29	276

Sources: Statisztikai Evkonyv and Mezogazdasagi Statisztikai Zsebkonyv, 1960 and 1975.

Fertilizer prices were increased by 23 percent in 1976 and consumption dropped seriously to 254 kilograms per hectare. Radio Budapest stated on August 31, 1977, that on some cooperatives, application barely reached 100 kilograms per hectare. Without question, price, not availability, is the decisive factor, and farms appear to favor running the risk of lower productivity than spending more for fertilizer.

The 1979 Plan called for a gross fertilizer consumption equivalent to 370 kilograms per hectare. But this seems excessively high and optimistic in a year when the emphasis in the Hungarian economy is on frugality if not outright austerity. It is unlikely that the state farm agency, Agrotroszt, will be able to deliver more than 80

percent of this amount. Cost factors are important
because the Hungarian fertilizer industry is able
to produce only about 55 percent of the domestic
requirement; 30 percent is obtained from other
COMECON countries; the balance comes from
hard-currency areas, where production costs and
prices have risen substantially, chiefly as a
result of the rapid rise in petroleum costs. It
does not appear likely that the fertilizer requirements of an expanding CPS program will be met for
some years.

Although shortages of machinery and fertilizer
are the two major elements, limits on investment
capital for other requirements also inhibit CPS
growth. These include specially tested hybrid
seeds, for which there is no Hungarian production.
All must be imported from the West, chiefly from
the United States. Herbicides and pesticides essential for efficient CPS operation are either imported or produced locally from imported chemicals,
usually from hard-currency areas.

Another way of evaluating the CPS is through
the results of the livestock program. Although
there have been very substantial exports of grain
from increased CPS productivity, a basic measure of
rural transformation is in domestic consumption
levels of meat, which the government has consis-
tently insisted will and must grow substantially
higher. Yet, the supply of meat for the Hungarian
people has been inadequate, often irregular, and
remains a chronic problem. In 1978, however, the
total amount of meat slaughtered increased 4 percent over 1977 and the amount of meat sold in shops
rose 4.7 percent. (12) Meat consumption has been
increasing, but lags well behind that of western
industrial countries. It reached 73.3 kilograms per
capita in 1978, and is expected to increase in
1979.

Increased exports of meat to both Western and
COMECON markets have cut sharply into supplies
available on domestic markets. It appears unlikely
that exports to the West will be curtailed to
increase domestic supplies, because export profits
are high and more hard currency is required to
reduce foreign trade deficits.

An interesting and significant comment on the
overall effectiveness of the CPS in meat production
is the fact that 55 percent of all slaughter hogs
in 1978 came from the private sector - from private
plots and the small number of private farms remain-

ing. (13) Recognizing the importance of the private
sector in farm production, the government has long
since given up serious efforts to eliminate this
"un-socialistic" element. Greater productivity is
encouraged in the private sector, and on some farms
there is substantial direct support to improve
private plot output.

The government is committed to substantial
growth of the capital-intensive CPS, but it is
either unable or unwilling to supply all the
capital required. It continues to depend strongly
on the heavily labor-intensive, small-scale, private sector for sufficient food, and is thus on the
horns of a dilemma - a dilemma which is unlikely to
be resolved, because of the delicate balance between ideology on the one hand and the absolute
need for productivity on the other. This is the
irony of the situation. Clearly, a strong case can
be made for substantial support for the private
sector as a realistic means of improving overall
agricultural output. It might even be that a
proportionate amount of capital, appropriately invested in the private sector rather than on the
CPS, would provide greater returns in terms of
improved domestic diets as well as exports. But
such a course would surely impale the regime on the
ideological Furthermore, such a commitment would
probably slow rural transformation as well.

Even with the CPS, the recent growth pattern in
Hungarian agriculture shows dependence on two intensive production extremes: the capital-intensive
CPS program, which now occupies about one-third of
the total crop acreage of the country, and the
private sector, heavily labor intensive, which
appears to come closer to the real potential of the
land than has as yet been achieved by the CPS.

It must be suggested that there appears to be a
less than full commitment by the government to the
CPS. An interesting example of this has been a
small change in terminology used in connection with
the means and goals of the CPS. Initially, it was
stated that "the most modern" machinery and equipment would be obtained and used in the program. By
1977, however, the term most commonly used to
describe the same aspect of the program was "the
best obtainable" machinery and equipment. This
implies that the glow may be off the program, and,
while the CPS continues to grow and cooperation
with American firms continues, the CPS, function-
ally, has reached a sort of plateau. Future growth

appears unlikely to continue at levels of the earlier years. (14)

It is not clear what has lead to the slowing of CPS development. Some reports imply that county and district party officials have complained that their local influence has declined because the CPS gives the farms too much autonomy. Whatever the cause, there is a slowing of CPS development as the final year of the current five-year plan approaches. (15) In spite of the substantial capitalization of agriculture through the CPS, the industrial sector has continued to grow rapidly (Table 15.3).

Table 15.3. Contributions to National Income
             (in percent)

	1960	1975
Industry	36	45
Agriculture	29	15

Sources: Statisztikai Evkonyv, Mezogazdasagi
          Statisztikai Zsebkonyv, 1960 and 1975.

Because the CPS has sought, with substantial success, to duplicate corn-belt agriculture, a comparison of areas with similar physical conditions can give some measure of the Hungarian CPS potential. The analogue area method of comparative regional analysis between European and North American areas has been well tested and offers a means of comparing productive systems by eliminating the variables of the physical environment (16).

While it is desirable to obtain exact physical analogues of climate and soils similarities, as well as of drainage, elevation, and latitude likeness, the latter are not fully realizable in a comparison of Hungary with the United States. But temperature and rainfall data for Hungary compare closely with those of counties in north-central Kansas and south-central Nebraska, with 50 to 70 centimeters average annual precipitation. There are however, marked differences in elevation, drainage, and latitude.

The counties are at 500 to 600 meters above sea level, while much of Hungary lies below 200 meters. Drainage can be slow and difficult, and flooding is common in parts of low-lying Hungary. Latitude affects hours of daylight during the

growing season, an important factor in crop production; the Nebraska and Kansas counties lie at 40
to 42 degrees north, while Hungary is mostly
between 46 and 48 degrees north. These counties
are at the southwestern limit of the corn belt and
the northern perimeter of the winter wheat region -
both, therefore, are marginal in terms of optimum
crop productivity. Although Hungary has adopted the
technology of the best of the corn belt, its
natural environment is distinctly inferior to the
areas where the technology was developed.

The purpose here, however, is not to consider
the corn belt analogue in detail, but rather to
view it as a measure of the potential of Hungarian
agriculture with continued and consistent operation
of the CPS. Development toward this potential would
be the basis of significant rural transformation.
Accordingly, an average has been calculated for the
twelve-county area from data in the 1974 census of
agriculture (Table 15.4).

These averages show the differences as well as
some marked similarities between the CPS in Hungary
and the Nebraska-Kansas (N-K) analogue. United
States studies of optimum-sized corn belt farms,
considering crops, capital inputs, machinery size,
farm labor, and marketing, suggest that the optima
are between 160 and 200 hectares. The much larger
Hungarian "units" may well result in reduced efficiency and lower returns. Under the cooperative and
state farm structure of the Hungarian system, which
has shown a strong trend toward larger and larger
farms, these sizes perhaps can be justified.

There are obvious management difficulties for
the CPS, whatever the size of the operating unit,
because of the cooperative and state farm organization, difficulties that do not exist for the
primarily family owned and operated farms of the
N-K analogue. It is the mixed character of the CPS
program on only parts of the larger farms, while
the other fields follow older farm methods, which
creates managerial problems. One system cannot help
but affect the other, directly or indirectly, so
that neither can operate most effectively. It must
be presumed that, despite present slowing of
growth, ultimately all areas of all farms will be
CPS. This means that the greatest need at this time
is essentially horizontal CPS development - an extension of CPS over all parts of the farms using
CPS - as a first step. Then the minority of farms
not using CPS can be brought into the program. In

Table 15.4.   Nebraska-Kansas (N-K) Analogue
             (12-county area)

N-K area	
Population Density, 1970	5.9 per square kilometer
N-K area	
Rural Population Density, 1970	3.7 per square kilometer
N-K area	
Number of Farms, 1974	10,289
Average Farm	
Size, 1974	196.4 hectares
Average Farm	
Value: Land, Buildings, Machinery	$213,716
Average Farm	
Total Value per Hectare	$1,088
Average Farm	
Hectares per Tractor	88.1 hectares
Average Farm	
Wheat Yields, quintals/hectare	20.7 quintals
Average Farm	
Corn Grain Yields, quintals/ha.	59.9 quintals
Average Farm	
Number Cattle/Calves	75
Average Farm	
Number Hogs	143
Average Farm	
Value Farm Products Sold	$55,055
Average Farm	
Value Crops Sold	$27,648
Average Farm	
Value Livestock Sold	$24,243
Average Farm	
Other Sales	$3,164
Average Farm	
Production Costs	$34,980
Average Farm	
Sales Minus Costs	$20,075
Average Farm	
Sales Minus Costs per hectare	$102.20

Sources: U.S. Census of Agriculture, 1974, U.S.
          Census, 1970.

area terms, these steps mean an increase of some 200 percent, with proportionate but probably higher capital inputs as a result of inflation and higher energy costs.

The greatest difference beetween the analogues--one which the CPS can hardly change rapidly, but one remaining at the root of Hungary's agricultural problem--is the great variation in rural population densities. Farm areas of Hungary have rural densities averaging eight to ten times those of the N-K analogue. Of course, by any European and even most Soviet rural densities, the N-K density of 3.7 per square kilometer is almost shockingly low and inconceivable in terms of European agricultural and social history. Dense rural population is perhaps the most important factor working against the success of the capital intensive CPS, and remains a significant constraint on rural transformation.

Although Hungary has continued to lose rural population for many years, (Table 15.5) rural densities of even ten per square kilometer, well over twice that of the N-K analogue, must remain decades away.

Table 15.5. Persons Working In Industry and Agriculture

	Numbers		% of labor force	
	1960	1975	1960	1975
Industry	1,384,000	1,816,300	29.3	35.7
Agriculture	1,849,000	1,156,800	39.3	22.3

Sources: Statisztikai Evkonyv, Mezogazdasagi Statisztikai Zsebkonyv, 1960 and 1975.

Any anticipated rural transformation must take this long-term fact into consideration. Thus, it can be predicted that a substantial rural transformation in terms of greatly improved incomes and levels of living must remain a long-range goal rather than a condition to be realized soon. Even if there were any assurance of urban employment, income and housing for many of the present rural Hungarians, they cannot at present be spared from the farms because their labor continues to be needed.

It may be argued that making Hungarian agriculture more capital intensive by the CPS may not
be the most effective way of reaching desirable
goals of rural transformation in the near future.
One cannot help but note the absence of any
substantial steps toward increased irrigation in
the CPS program up to the present. The prospects of
greatly expanded irrigated lands in those areas
well suited to such development as a means of rural
transformation should be considered. To be sure,
irrigation agriculture is also capital intensive.
(17) But the same capital inputs going into irrigation, rather than the CPS, would not be quite so
likely to result in the large amount of labor
redundancy that is certain to take place through
the CPS, especially when farm efficiency approaches
that of the N-K analogue. Labor inputs for row
crops and other products raised under irrigation
would continue to be high.

In sum, it is clear that the Hungarian potential for achievement of the productivity levels of
the N-K analogue through CPS operation in the
relatively short run is possible, but not likely.
The government's enthusiasm for further expansion
of the CPS has diminished under the pressure of
dwindling hard-currency reserves and an unfavorable
balance of payments. But without the continued
capital inputs, which can come only through the
government's commitment, the sizable gains under
CPS to date can mean that Hungarian agriculture
will be able to maintain these moderate productivity levels. The prospects for significant rural
transformation remain limited in the short run,
though the potential for the longer run is high.

NOTES

(1) A. Elias, "Contributions: Third National
Conference on Agricultural Economics," Bulletin
No. 42 (Budapest: Research Institute for Agricul-
tural Economics, 1978), p. 93.

(2) Details of the modifications of the Soviet
models may be found in F.E. Dohrs, "Incentives in
Communist Agriculture: the Hungarian Models,"
Slavic Review 27, no. 1 (March, 1968): 23-38.

(3) Nepszabadsag, Nov. 18, 1971.

(4) Although there are several acronyms applied to the various closed production systems, CPS for Closed or Crop Production System is the most useful inclusive term for all variants now operational in Hungary.

(5) Radio Budapest, December 4, 1972.

(6) Magyar Nemzet, Oct. 2, 1976.

(7) Nepszabadsag, Dec. 24, 1978.

(8) Magyar Hirlap, Jan. 7, 1977.

(9) Kisalfold, Dec. 3. 1976.

(10) Vilaggazdasag, April 6, 1979.

(11) Magyar Statisztikai Zsebkoyv, 1977.

(12) Magyar Nemzet, Feb. 14, 1979.

(13) Szovetkezet, Nov. 8, 1978.

(14) Magyar Mezogazdasag, Jan. 5, 1977.

(15) Janos Fenyvesi, "Contributions: Third National Conference on Agricultural Economics," Bulletin No. 42 (Budapest: Research Institute for Agricultural Economics, 1978), p. 110.

(16) A number of studies have been made using the analagous area method, including two which have been published: S.S. Birdsall, "The Effect of Management on Crop Yields in Soviet Agriculture," Journal of Geography 67, no. 2 (February, 1968): 95-103; C. Zoerb, "The Virgin Land Territory: Plans, Performance, Prospects," in Laird and Crowly, Soviet Agriculture: the Permanent Crisis (New York: Praeger, 1965), p. 29-43. Other students under the writer's direction have completed similar research on Romania and Yugoslavia and their North American analogues.

(17) Fred E. Dohrs, "Nature Versus Ideology in Hungarian Agriculture: Problems of Intensification," in G. Hoffman (ed.), Eastern Europe: Essays in Geographical Problems (London: Methuen, 1970), pp. 271-300.

IV
The Industrialization of Agriculture and Villages

16 Process and Response in the Industrialization of Agriculture

Michael Troughton

This chapter discusses a general model of the processes of and responses to change in modern agriculture. The model is descriptive of the dynamic relationships between agriculture and industry in developed countries.

CHANGE IN MODERN AGRICULTURE

World agriculture has undergone greater change in the last thirty years than in any commensurate period in history, and rural transformation has occurred on an unprecedented scale. However, while recent events such as food and oil crises emphasize interdependence within a global system, most agricultural change is taking place independently within the distinct subsystems of developed and less developed countries. Furthermore, although agricultural change has its greatest impact through transformation affecting billions of peoples in the predominantly agricultural Third World, the momentum of change is most rapid in modern agriculture, which is virtually confined to developed industrial societies.

The concept of change in modern agriculture for which a generalized model framework is here attempted, is centered on the so-called industrialization of agriculture. Modern agriculture, which developed in conjunction with, but remained operationally distinct from industry, is now commonly experiencing changes which seem set to result in

213

the absorption of a distinct agricultural-rural sector of society into the dominant urban-industrial mould. Agricultural industrialization represents a situation in which the basic input-output model of modern agriculture is being amplified by a process-response sequence based on the adoption of common agro-industrial technology and economic ideology, expressed as a set of common operational processes; increased scale, intensification, specialization, and integration of farm production (Figure 16.1).

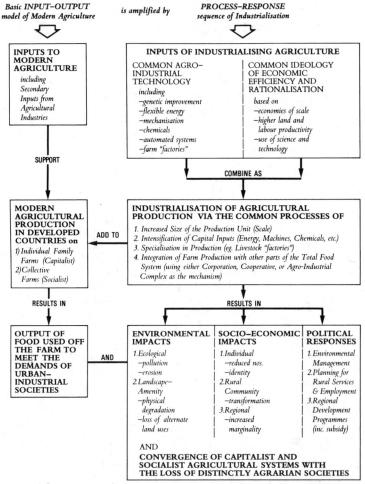

Figure 16.1. Model Framework for Industrialising Agriculture in Developed Countries

Somewhat paradoxically, this viewpoint is an outgrowth of recent contact with agriculture in six quite differently developed countries: Canada, Denmark, Finland, Great Britain, Hungary, and Poland. While contact confirmed that physical, socio-economic, and ideological differences do contribute to distinctive national and regional agricultures, it also provided evidence of the common phenomena of industrialization. The processes of industrialization seem to be common to both capitalist and socialist agriculture and, furthermore, seem to result in a common set of responses. The latter include generally unwelcome impacts on both the natural environment and human community, and complex political responses (Figure 16.1). It seems important to try to generalize the nature of change affecting modern agriculture because, although agriculture affects fewer people directly in developed than in less developed countries, the change is to types of agricultural system to which many less developed countries aspire. Furthermore, the changes to agriculture threaten to radically alter, or even eliminate, distinct rural societies.

INDUSTRY AND MODERN AGRICULTURE

The successful development of modern agriculture in Western Europe, North America, and Australia reflected a close relationship with the course of the industrial revolution. While key changes in farm structure and operation, including the shift from common to individual tenure and the introduction of fodder-crop and rotation systems, were generated within farming, the predominant agrarian model that developed of commercially oriented agriculture practiced on individual family farms required both the demands and support of industry to achieve extensive adoption. Following the shift by manufacturing from cottage to factory scale, the dynamics of change became the urban-industrial demand for increased agricultural output and surplus rural labor, including the use of emigrant settler populations who produced food for metropolitan markets using new industrial products to facilitate extensive farming, transport, and processing (as in Canada).

Modern commercial agriculture developed side by side with industry during the nineteenth and early twentieth centuries. However, although growth was

mutually stimulating, strong operational and ideological differences were maintained. Industry, after an initial, dispersed, "cottage" stage, became
a predominantly large-scale factory operation using
large inputs of labor, fuel, and machines at
specific locations. Agriculture, despite a market
orientation, actually proliferated through the establishment of greater numbers of relatively small
family units; dispersion and "cottage scale" represented major distinctions in the nature of the firm
between agriculture and manufacturing. In the
twentieth century, manufacturing adopted the integrated assembly line and achieved new levels of
concentration and specialization, whereas agriculture, despite a gradual adoption of mechanization,
attained its greatest efficiencies through
small-scale, intensive, mixed operations (as in
Denmark). Although industrialization was applied to
the articulation of agricultural output (for example, the system of rail movement, elevator storage,
and milling of prairie grain), most agricultural
processing remained local and small-scale in the
prewar era.

As significant as the scale and operational
differences between farm and factory was the growth
and prevalence of the belief in the separate nature
and values of the urban-industrial and rural-agrar-
ian sectors of capitalist society. Agricultural
transformation in Great Britain involved great
hardships for an oppressed rural laboring class.
But elsewhere, greater emancipation (as in Denmark)
or settlement based on pioneering (as in Canada)
fostered a strong attachment to the concept and
reality of a class of individual freehold farmers
and an associated rural society serving the farm
community and sharing its values - values which, in
contrast to those of the urban proletariat, were
regarded as inherently virtuous and fundamental to
society. In countries where the agricultural sector dominated (for example, Denmark, Finland, and
western Canada), agrarian populism demonstrated
strong political influence. Even after 1945, pioneer settlement was actively encouraged in Finland
and Canada, while in Denmark and Finland legislation was enacted to limit increases in farm size
and maintain small family holdings.

Although this agrarian structure seemed both
stable and efficient as late as the 1950s, it had
already begun to be undermined. The primary agent
of change was the process of industrialization

advanced under the banner of progress in both technological and ideological contexts.

TECHNOLOGY AND IDEOLOGY

Industrialization in agriculture is associated with the postwar period. But some important contributing events occurred before 1945, notably, the Russian revolution and subsequent collectivization of agriculture, the economic depression of the 1930s, which forced a drastic revision of the captialist system, and the beginnings of the integrated application of science and technology at the farm level.

Collectivization, although not influential outside the USSR before 1945, represented a major alternative model of agriculture in which the organizational parallels with industry were made explicit. It was on this basis that it was adopted as the general model for socialist agriculture in the postwar period. The Depression affected all segments of the capitalist society, especially the extensive, recently settled farming areas (for example, the Prairie Provinces), indicating the weakness of individual small farmers. Of particular importance were the processes involved in economic recovery, whereby a general emphasis on government intervention and centralized planning challenged the individualistic ethic of capitalist agrarian society (as in Canada) and strengthened moves to integration based on cooperation (as in Denmark and Finland) or other types of rationalization.

Although the economic depression of the 1930s hindered adoption of agricultural innovations, there continued to be a steady increase in the application of science and technology to farming elements and operations; improved plant and animal breeding, new machines, and the first agricultural chemicals were all available by 1940. Wartime pressures for greater production and massive labor substitution provided the stimulus to their adoption and to the beginnings of the dominant trend to capital-intensive means to increase agricultural production.

It might be argued that the postwar period has been marked by a fundamental division within modern agriculture, namely, between capitalist and socialist systems, based on opposing political ideologies

and the sharp distinction between communal-collec-
tive and individual-freehold tenure. However, examination of the economic and technological directions taken reveals what, in fact, seem to be
convergent processes of and responses to change
within the common framework of urbanizing, industrial societies.

The linkages between industrialization and agricultural development have been made most explicitly in socialist countries. In the USSR and the
countries of Eastern Europe that became socialist
after 1945, agriculture had tended to be technologically backward and organized around large peasant
populations. In these circumstances, collectivization to achieve modern agriculture represented a
close combination of political and economic goals:
the transformation of a rural peasantry not into
individual entrepreneurs, but into a rural proletariat sharing common socialist values with urban
workers; and the consolidation of peasant plots
into large units suitable for intensive capital
inputs and specialization, both of production and
of the reorganized rural labor force. The large
collective farms were the rural equivalent of urban
state-manufacturing enterprises, the basis for inputs (capital, machinery, chemicals) supplied by
state financial and manufacturing concerns, and
with output planned to meet the requirements of
state food-processing and distribution systems.

Although the key to socialist agricultural
organization may seem to derive from its political
ideology that seeks, above all, to eliminate individual, captialistic entrepreneurial activity, this
has been of much less significance than the base
provided by collective farms for the application of
large-scale agro-industrial technology. The large
consolidated land units have facilitated the adoption of large, specialized field machinery, heavy
applications of fertilizers and pesticides, and
specialized production systems, including animal
husbandry based on large herds of housed livestock.
Significantly, most of these innovations were developed first within capitalist commercial agriculture. However, not only do the collectives provide
the base for large-scale production technology;
they also fit the concept of a unified socialist
economy based on the efficient and integrated
operation of each economic sector. In this respect, they are well adapted to the application of
the principles of economic efficiency based on

economies of scale and integration of farm production with other parts of the total food system.

Capitalist agriculture presents a somewhat schizophrenic ideological outlook, with a marked ontrast between what may be termed political and economic dogmas. Political parties from conservative to social democratic include preservation of the family farm and rural way of life in their platforms, and the issues generate strong emotions among both rural and urban constituents. On the other hand, the dominant direction of society, and of much stated policy, is toward an economic viewpoint similar to that in socialist countries: economic rationalization based on the bureaucratic application of technology designed to give increased productivity in all sectors. All governments have policies that encourage the development of agricultural technology and its widespread application according to well-established principles of economies of scale, product specialization, and integration. Land consolidation, increased intensity of application of capital and energy, and large-scale animal husbandry have all been encouraged as desirable innovations for both the farm and general populations, specifically to increase individual farm incomes while keeping down food costs. Despite much rhetoric, little has been done to stem the results, in terms of drastic reductions in numbers of farms and farm employment. Even in countries where farmers have received economic maintenance support (as in Finland and Great Britain) or operate within nationally organized cooperative supply and marketing frameworks (as in Denmark and Finland), many farmers have felt the pressure of relatively low incomes and the cost-price squeeze. In countries where farm inputs and outputs are dominantly in the hands of oligopolistic private agribusiness and where the economic efficiency/business model is applied at the farm level without subsidy or maintenance programs (as in Canada), the direction of economic rationalization is even more pronounced.

PROCESSES AND EVIDENCES OF INDUSTRIALIZATION

The outcome of the convergent viewpoint of modern agriculture, based on the common belief in agro-technology and an ideology of economic efficiency, is seen in the industrialization of agricultural

production via a closely interrelated set of processes: increased scale of the production unit,
intensification of capital inputs, specialization
of production, and sectoral integration of farm
production within a larger food system (figure
16.1). Evidence of each process, albeit within
differing policy and operational frameworks, can be
found in the agriculture of all developed countries.

Increase in the scale of farm operation, as
with manufacturing, is closely related to a reorganization of the units of production; specifically, a decrease in the number of farms. Fewer
farms means an inevitable decline in the agricultural population, while consolidation provides the
framework for further labor substitution by means
of concentrated capital investment. Capital intensification is made through more efficient energy
sources, applied through machines to both field and
barn operations, and by the use of chemical fertilizers and pesticides, the array of structures to
house and feed large herds of livestock, and
increasingly through provision of automated operating and monitoring systems.

Larger scales of operation and capital intensification are facilitated by (and, in turn, encourage) specialization, both of a narrower range
of crops, each with its specialized machinery, and
of single types of livestock under "factory farming" conditions. Individual farmers specialize to
simplify decision making and streamline investment;
this strategy is mirrored by the managerial and
operational divisions aggplied on the larger, col-
lective-farm units.

Increases in scale, capital intensity, and
specialization seek to achieve the goals of increased productivity per unit of land, labor, and
capital invested at the farm level. On this base,
it becomes increasingly possible to achieve further
efficiencies through integration, both at the farm
level and between it and other parts of the
food-production system. In effect, the modern specialized farm unit, whether individual or collective, assumes the role of a "station" on the
food-production assembly line. To the station come
inputs of capital in the form of energy, machines,
and chemicals, and also scientific and managerial
expertise. From it flow large quantities of uniform output of limited range, which is the ideal
input to food processing. Organization to achieve

this level of integration may involve corporate, cooperative, or state interests. But the technical nature of the operation, its capital requirements, and therefore the measure of "corporate" concentration of control tends everywhere to be very similar. Insofar as the farm-supply and food-processing sectors are already industrialized, then integration tends to centralize control of agricultural activity outside the agrarian context.

Increases in farm size, reductions in farm labor force, and intensification and specialization of production are widespread (table 16.1). Although there still exists a tremendous difference between the scale of collective farms in Hungary and family farm units in Finland and Denmark, the rates of change are indicative of the general nature of the process. Hungary provides what is perhaps the best example of the planned operation of the overall set of processes, with development based on enlargement to create so-called agro-industrial complexes, whereby the production of several collectives is integrated to supply a complex of food-processing units. Further integration may involve development of common power facilities, specialist services, and even the negotiation of export supply contracts. The state sector in Poland is adopting a similar approach. The model is acknowledged in Hungary as derived from western experience, using energy-intensive production systems developed in the American midwest. Similar types of integration, but based on national cooperative marketing agencies, are found in Denmark and Finland. Although Scandinavian cooperatives are based on individual farmer membership, and in the past have sought to maintain production on the largest number of farm units, there is evidence that they now favor the elements of agricultural industrialization. Policy in both countries is aimed at consolidation of farm units; Denmark has experienced one of the most rapid rates of decrease in farm numbers, while Finland has instituted a number of schemes to encourage reduction in numbers. In both countries, there has been a marked increase in specialization that, together with increased size of farms and reduced farm population, has been accompanied by massive investment in machines and related inputs. Indicative of the direction being taken has been the rapid consolidation of cooperative dairy and slaughtering operations.

In Canada and the United Kingdom, where farm

sizes were already large, farms have continued to grow fewer and larger, and the farmer and farm labor groups have assumed miniscule proportions in the national workforce. Substitution has been by major investment in technology, and in both countries most livestock production is now concentrated in very large "factory" units. Although control of agriculture is still nominally in the hands of individual farm owner-operators, more and more production is based on contracts and financing arrangements controlled by large-scale, oligopolistic agribusiness. In addition, there is a high level of government bureaucratic intervention that has a strong influence on the types and disposal of production.

RESPONSES TO INDUSTRIALIZATION

Conventionally, the results of industrial activity are evaluated in the economic terms of output and/or productivity. Industrialized agriculture has been justified primarily in such terms. While there is considerable evidence of potential for increased output in absolute measure, and this remains a goal in socialist countries, much industrialization has been of modern agriculture facing a condition of inelasticity of demand for food. This means that the chief measure of economic performance has been of increased productivity of a relative type, that is, higher output per farm or farmer, but with fewer numbers of both. In economic terms, then, agricultural industrialization is most commonly resulting in rationalization rather than expansion of production. While rationalization may be justified in terms of economic efficiency, the results of agricultural industrialization are not confined to the economic sphere. Rather, as the final part of the model suggests, the operational processes result in a broad set of responses within the overall rural-agricultural system. Once again, the evidence suggests a common pattern, here generalized under three headings: environmental impacts, socioeconomic impacts, and political responses (Figure 16.1). These impacts are important because of their common, widespread, and problematic nature, which suggests that the industrialization of agriculture may be having a generally destructive effect on agrarian environment and society and may be necessitating a series of political responses that may or may not prevent the loss of those societies.

Environmental Impacts

From its inception, modern agriculture began to alter traditional man-land relationships to interrupt local cycling of materials and to introduce new techniques capable of greater environmental impacts. Industrialization, however, has vastly increased the scale and pace of such change.

Environmental impacts may be distinguished as ecological and landscape-amenity effects, but both are closely connected with the major processes of increased scale, intensity, and specialization of farm operation. Ecological impacts take place both on the cultivated and on the remaining "natural" ecosystems. The remnants of the latter are falling victim to widespread farm and field enlargement, designed particularly to facilitate increased scale of mechanization. The use of heavy machines and intensive, chemically supported grain cultivation each contribute to an increased potential for damage to soil structure and problems of loss of soil fertility and/or wind and water erosion. Runoff including fertilizers contributes to water pollution, eutrophication, and further damage to natural ecosystems. Similar effects accrue from the problems of waste disposal from large herds of housed livestock.

Many traditional and even recent agricultural landscapes contribute to what is regarded as rural landscape amenity, including farm and field layout, buildings of traditional style and materials, and the visual presence of livestock. The effects of industrialization, including farm amalgamation, field enlargement, monocultivation, housed livestock, and construction of utilitarian "industrial" farm buildings, contribute to a loss of amenity including a significant visual transformation of the rural milieu. In addition, there are the problems of air (odor) and water pollution and an increasing propensity for farming organized along agro-industrial lines to produce landscapes unsuitable for alternative countryside uses, including both outdoor recreation and nature conservation.

Environmental pressures are being identified in all developed countries, although they are greatest in relation to the scale of the farm operation. Thus, collectivization is creating a major impact

on the landscape through its huge consolidated
fields and concentrations of housed livestock and
other structures in the open countryside (as in
Hungary and Poland). Field enlargement and vegeta-
tion removal are concerns in England and Canada,
while problems of water quality, runoff, and odor
are widespread. Both Denmark and Finland have
experienced major landscape changes and problems of
preservation of the natural environment.

Socioeconomic Impacts

Agriculture is, above all, a human system, and
socioeconomic impacts are of fundamental importance. These too have increased in scale and
intensity as a result of industrialization. The
major distinctions are between impacts affecting
the farm population, those affecting the function
of rural communities, and broader impacts on the
viability of agrarian society at the regional
level.
    Industrialization generally means a drastic
change in the way of life of the farm population.
Many have been made surplus and been forced to
abandon farming; many others have taken up off-farm
employment and operated as part-time farmers. In a
capitalist setting, the individual farmer, to survive the competitive pressures, has to adopt a more
purely business approach. This carries major pressures in terms of technical and financial responsibility. Fewer numbers of farms and increasing
specialization of farming tend to isolate the
farmer, even within rural society. This is less so
in cooperatively based societies (such as Denmark
and Finland), but here too, there are problems of
loss of individuality. The latter may be a particular problem in socialist agriculture where the
majority of persons operate as farm workers and
where even the variegated nature of traditional
peasant tasks has been reduced.
    Farmer, farm family, and farm worker impacts
contribute collectively to change in the rural
community. Rural-settlement functions change in
response to the overall reorganization of agricultural supply and processing. Increased scale and
integration has resulted in major reductions in the
size and number of rural settlements in capitalist
countries, with widespread concern over problems
not only of maintaining population but of providing

basic services and amenities, including employment, retailing, education, health, and transportation (as in Finland and Great Britain). Collective organization may better preserve the rural community, especially where it is based on relatively large rural settlements (as in Hungary), but here too, widespread adjustments have to be made, including provision of alternative rural employment.

A third major impact is that of increased regional differentiation, specifically the problem of agriculturally marginal areas. By its nature, industrial agriculture emphasizes physical and other locational advantages. Consequently, a universal by-product is the increased relative disadvantage experienced by poorly endowed, low--productivity, and isolated areas. In each country, viable agriculture is concentrated on a smaller area, and marginal areas are enlarged, giving rise to social and economic hardship and problems of regional disparity. This applies equally to poorer areas in maritime and "near north" Canada, highland Britain, central and northern Finland, or highland Hungary. In Poland, the problems of mountainous areas are exacerbated by the small-scale, fragmented nature of the private sector.

Political Responses

Agricultural industrialization is taking place in an era of increased government activity and increased public concern. Nevertheless, it is contributing to the need for even more political response, particularly in the areas of environmental management, planning for rural services and employment, and the need for regional development programs.

The nature of agricultural practice, including chemical inputs, constitutes a major concern for the government ministries responsible for the environment, including the need to formulate regulations to control air and water pollution, to preserve rural amenity, and to resolve areas of conflict involving agriculture and other rural land uses.

The responses to the rapidly changing socioeconomic situation are many. In socialist countries, planning for rural employment, services, and amenities is regarded as part of overall regional planning; it involves attempts to increase rural employment, income, and opportunity, and to upgrade rural services. In some capitalist countries (notably Finland), a similar comprehensive approach is taken, including integration of measures to rationalize farming (with retirement schemes and land banking, for example) and those to maintain rural community structure through education, health, and transportation facilities. In Canada and Great Britain, especially, the focus is on regional economic disparity, with attempts to provide alternative employment to replace or enhance agricultural income. Nevertheless, the paradox usually remains of parallel measures to encourage industrialization and thus accentuate the distinction between the suitable and unsuitable areas for its application.

CONCLUSION

While different conditions of physical base, the backlog of social and economic experience, political ideology (and expedience), and considerable sentiment will continue to differentiate between the evolving agricultural systems of developed countries, an attempt has been made to indicate a high degree of convergence at the present time in the direction and resultants of industrial change. The process-response model advanced suggests the operation of a common set of processes: increase in scale, intensification, specialization, and integration. Considerable evidence exists of the operation of these processes and of a set of similar responses, especially impacts on environmental and socioeconomic systems. Most interesting is that industrialization of agriculture is occurring in both socialist and capitalist countries. This might suggest that the most powerful forces shaping agriculture are commonly available types of technology and their technocratic application. Another conclusion might be that political ideologies are less significant than a general bureaucratic thrust toward economic rationalization in agriculture. Finally, rural transformation, about which one tends to think in essentially human terms, appears as the by-product, or residual, rather than as the raison d'etre for agricultural change, and this is perhaps the danger of the overall situation.

ACKNOWLEDGEMENT

Grateful acknowledgement is made of a NATO Challenger of Modern Society Fellowship and a Finnish Government Scholarship which enabled comparative study of agricultural industrialization to be undertaken during 1979.

17 The Effect of Modern Agriculture on the Development of Rural Settlement Network: The Case of Hungary

Gyorgy Enyedi

MAIN CHARACTERISTICS OF RURAL TRANSFORMATION IN HUNGARY

Prior to World War II Hungary was an agrarian-industrial state with a poorly developed industrial base. According to the census of 1941, 50 percent of the active population was employed in agriculture and 60 percent of the total population lived in the rural settlements. As a result of the speedy urbanization and industrialization since 1945, today less than 20 percent of the population is employed in agriculture and 51 percent of the population lives in the cities. Since our economic geography and settlement science specialists have also been primarily concerned with this rapid urban-industrial development, there was scant attention paid to research concerning the rural area. The development of the settlement network and regional planning was also urban centered. Consequently, the inadequately planned and often spontaneous transformation of the rural settlement network is not sufficiently known.

The importance of the rural sphere, however, can scarcely be doubted. 49 percent of our population and 50 percent of all industrial workers still live within it, and the total population of the rural area decreased by only 3 to 4 percent during the last four decades. The importance of agriculture - the primary function of the rural sphere - has increased as a result of the development of extensive foreign trade activities and of the strategic and economic role of food production.

229

The definitions of "rural" and "urban" sphere vary from country to country; in general, they reflect the historical development of the settlement network of a given state. In Hungary, the village and the city are primarily administrative units, and the statistical concepts of "urbanity" merely follow the administrative lines. The size of the population is not a primary criterion: in the Hungarian Great Plain there are rural settlements (villages) with fifteen to twenty thousand people residing in them. Generally, a settlement becomes officially recognized as a city considerably later than the functional existence of urban networks. Administratively, there are only 98 cities in Hungary, even though, according to the findings of settlement geographers, 140 to 150 settlements fulfill urban functions. If one were to employ the same definition as that which is used in the GDR, 80 to 82 percent of Hungary's population would be identified as "urban". Hence one must be careful when one utilizes cross-national comparisons.

In the Geographical Research Institute of the Hungarian Academy of Sciences, research has been concentrated on the transformation of the villages of Hungary since 1976. In this context, the rural sphere has been defined as that area which - along with its natural resources, settlements, and population - falls outside the urban agglomerations. Thus, even small towns existing outside such agglomerations are regarded as part of the rural sphere.

The general characteristics of the transformation of the rural sphere that has taken place during the last thirty years are discussed below.

The size of the rural population has decreased only minimally, even though within the rural sphere one could witness major population movements. Significant decreases occurred in the nonindustrialized tiny villages of southwestern Hungary and the northern mountainous region, where there are no major urban centers and where the natural agrarian environment is unfavorable.

At the same time, the population has increased in the commuting zones of the cities, especially around the northeast-southwest industrial axis and in those villages with more than five thousand inhabitants that fulfill some urban functions. The intrastate migrations have decreased, and such migrations that do take place are no longer unidirectional toward the cities. Between 1970 and

1975 the majority of permanent migrants moved from one village to another, and there has been labor movement among the various agrarian production units as well.

The socioeconomic structure of village and city have become similar to each other, although, naturally, only 10 to 12 percent of those employed in agriculture lives in the cities, and the percentage of those employed in the service sector is much higher there. It is significant that the ratio of industrial workers is the same in the city and the village. Unlike thirty years ago, when most people living in the village were agricultural employees, today only 38 percent of the rural population is employed in agriculture. Moreover, the gap between industrial and agricultural incomes has been eliminated during the last ten years. A fundamental feature of their social equalization has been the basic transformation of agricultural employment.

The previously predominant agrarian character of the rural sphere has become multifunctional. Especially between 1960 and 1975, significant industry moved into the villages. At present around one-fifth of the factory workers (around 400,000 individuals) work in factories located in the villages, and this figure does not include those working in the industrial production branches of the agricultural cooperatives. Even the residential function of the rural settlements expanded significantly; more than one million workers commute daily from rural residences to urban workplaces. In addition, the tourist and environmental-protection functions of the rural sphere have also been on the increase.

The modernization of rural function has been faster than the modernization of the rural settlement infrastructure. Consequently, tensions have been evident between the speedy socioeconomic development and the level of rural existence. On the one hand, progress was evident: by 1963 all of the villages had been supplied with electricity, by 1974 they were all connected with the national road and bus transportation networks; by 1978, 60 percent of the rural homes cooked with gas and 40 percent had running water. On the other hand, the level of rural existence was significantly below that of the city infrastructurally and in regard to services. Thus, in Hungary at the end of the 1970s, a chief source of social inequality remains the difference between "town" and "country".

The transformation of the Hungarian village took place as a result of exogenous factors such as the general process of urbanization, and indogenous factors, such as the general modernization of agriculture in Hungary. In spite of the decrease of the agricultural population, agriculture remains the most important rural function: 75 percent of the land remains under agricultural use and a further 15 percent under forestry use. The annual growth index of food production since 1970 is hardly below that of industrial production. According to the 1978 report of the World Bank, between 1967 and 1977 Hungary's agricultural growth had been the second in the world, exceeded only by that of the Netherlands.

THE CHARACTERISTICS OF MODERN AGRICULTURE

One can discern several phases in the development of modern agriculture. The first is the period of mechanization and economic concentration that creates the technical precondition for large-scale agricultural production. The second phase is the period of the so-called biological revolution that creates the preconditions for the large-scale, serially produced, similar quality produce through the introduction of higher plant and animal yields. The third phase is characterized by the fundamental modernization of economic organization, during which agricultural production is suffused with industrial production and merchandizing techniques. The characteristics of the agricultural transformation of the 1970s and 1980s is this organizational change.

The most significant aspect of this organizational change is the integration of previously sharply delineated economic activities. While in traditional agriculture the production of most of the tools necessary for that production (that of the produce and the sale of the end product as well) was done by the peasantry itself; today four groups of producers must harmonize their activities to insure the production and merchandizing of agrarian goods. These are:

1. The service branches that insure supplies and means of production

2. The so-called technical-intellectual branches that insure the development of such services as research, consulting, and specialist training

3. The agricultural production itself

4. The processing and sale of the final agricultural produce

Integration in developed socialist and capitalist large-scale agricultures is fundamentally different. In the capitalist system, through the integration process, the functions of the farm decrease, their operations more and more are directed from the outside, and the roles of the integrators are played by the large agricultural-commercial firms. The family farm is increasingly dependent on the industrial and commercial organizations existing above the level of the farm and becomes a cog in the wheel of production. The farmer remains independent, but control over and decisions about his own production activities slip out of his own hands. Thus, for example, the complete organization of poultry and egg production was previously in the hands of the farmer who had selected his stock, grew his chickens, and sold his eggs on the market. As a result of the integration process, the breeding the fodder, and the merchandizing has been taken out of his hands; the mammoth agricultural firms today decide on the proper type of chicken and egg hybrids, preparation of fodder and forage, and the sale of the eggs. The production of food has been diffused geographically as well as organizationally; storage, preparation, and feed prduction have been separated from the production locus and have been concentrated largely in the cities. In the capitalist agricultural-industrial integration, the comparative importance of agriculture has been sliding. In France in 1972, for example, the agricultural share of food products from the final price was only 28 percent, and the concentration of the large firms and multinational ones has been on the increase. In contrast, the integration of large-scale socialist agriculture has been characterized by other general processes. The large-scale farms, whether state or cooperatives, can conduct their activities on an equal basis with the commercial or

food-preparation firms. As a result of the small
size of the farm in capitalist systems, there are
very few farms (such as the King's Ranch) that can
afford to do that. Lest one forget, an averagesized
Hungarian cooperative is fifteen times larger than
an average-sized American farm. The large farms in
socialist systems create joint companies with the
state food-industrial or large-scale commerical
firms; the levels of integration become blurred,
but the agricultural unit does not become subordinate to higher levels.

The large-scale farms produce such enormous
amounts that they engage at least partially in
their preparation and sale. The development and
breeding of new branches and hybrids, the employment of new production processes, and innovation in
general are usually tied to state farms and to a
few special cooperatives, and therefore remain
within the agricultural production branch itself.
Hence, through the process of horizontal integration - for example, in the employment of various
production systems - various economic units of the
same level come into close contact with one another.

The processes described above have decentralized a large number of industrial, transportation,
and commercial activities in the rural sphere by
the large agricultural firms. As a result, a large
part of the population listed in statistical indices as employed in agriculture is, in reality,
engaged in nonagricultural work;

Finally, while integration in the developed
capitalist states results in a functional impover-
ishment and a consequent further loss of prestige
of the village, the large-scale economic activities
taking place in Hungary generally stabilize rural
life, broaden rural activities, and sometimes bring
dynamism to the locus as well.

THE EFFECTS OF MODERN AGRARIAN DEVELOPMENT

The technical-organizational change briefly
sketched above has naturally had a significant
influence on rural settlements. Here, attention
will be concentrated on the influence on the
settlement network itself, without regard to the
internal settlement structure of the village.

Internal Transformation
of the Agricultural Population

The concept of agricultural population completely
changed during the course of collectivization: the
old traditional peasantry nearly disappeared from
Hungarian villages. In the large-scale agricultural
units, traditional peasant labor - based on familial division of work and universal type of agrarian
activity - was no longer needed. The breakdown of
labor processes into smaller component parts made
technical adaptation easier, although, today's agricultural skilled laborer is less able to see the
fruits of his labor.

In 1973 the vast majority, 70.5 percent, of
agricultural employees worked in cooperatives; 15.2
percent in state farms; 5.1 percent in fisheries;
4.2 percent in forestry, and 5 percent worked in
their own private lands. The cooperatives also
undertook significant industrial activities: one-
third of their total income came from nonagricultural activities. Thirty percent of all cooperative members and employees were engaged in these
industrial, construction, commercial trade, or
warehouse activities. Moreover, of those listed as
being employed in agriculture, fully 12 percent was
engaged in white-collar, administrative or management activities. Thus, the roughly 18 percent who
are officially listed as being engaged in "agricultural" work cannot be compared on a one-to-one
basis with agricultural laborers engaged in private
economic activity.

A special characteristic of Hungarian agriculture is the enormous importance of auxiliary farming. In this type of activity, the traditional
familial division of labor continues to predomi-
nate; hence, the sociologists refer to the dominating life-style as that of the "post-peasants."
Although originally intended to serve the food
self-sufficiency needs of the cooperative peasantry, by now its importance has vastly exceeded the
presumed goals.

The majority of the household plots are owned
by nonagricultural employees: 40 percent of all
Hungarian families are engaged in these activities;
30 percent of all income derived from the household
plot is accrued to the working class, 23 percent to
the peasantry and the remainder divided among the
rest of society. The auxiliary farms and household
plots are more and more used to produce for the

plots are more and more used to produce for the
market, and by now more than 50 percent sell there.
In 1975, 36 percent of gross agricultural production value came from the household plots that
comprise only 14 percent of all agricultural land.
The household plots are engaged in the most intensive type of production activities: they offer 50
to 55 percent of all hog, poultry, viticultural,
fruit, and vegetable production. The household
plots have exceedingly close ties with the large-
scale farms; they buy their fodder and lease their
machines from them; thus, the household farm can be
regarded as an integral component of socialist
agriculture. The household plots save a large
amount of capital investment for the state through,
for example, the use of available stables, and also
utilize excessive labor. According to some calculations, 2.7 billion man-hours were spent on the
household plots in 1972-1973, only 30 percent less
than on the large-scale units. Of these 2.7 billion man-hours, however, 61 percent was worked by
those in retirement and by dependents.

Among the agricultural employees, those engaged in skilled labor and industrial or service
activities continue to demand greater services and
better technological, infrastructural environment
in the village. Since the consumption habits of
rural and urban residents are beginning to be
similar to one another, rural residents begin to
demand services similar to those found in cities.
The great importance of the household plots is that
they act as a conservative force: they retain at
least partially, the productive function of the
rural residence, and through this, maintain the
ground plan of the rural settlement as well.

New Ties Between the Large-scale Farms
and the Rural Settlements

The growth of farm-size had important effects upon
the rural settlements. During the 1970s the state
and cooperative farms were merged with neighboring
units in order to create more size-efficient farms:
by 1978 their numbers were reduced to 141 and
1,425, respectively. Consequently, in 1978 the
average area of a state farm was 7,043 hectares,
and that of a cooperative farm was 3,624 hectares.
Thus, on more than one-third of the territory of
the state - in areas characterized by small clus-
tered rural settlements - a farm area includes

several villages as well. While earlier a village
provided a frame for the agricultural activity, now
the farm holds together several villages, in some
of which there certainly are people who are employed in other sectors of the economy. The settle-
ment-development and economic-development plans are
not always sufficiently harmonized. Thus, for example, the unified cooperative does not always move
to the administrative center to share the centrifu-
gal role of the local council. The development of
the villages almost always is based on local
efforts and sources, and among the income sources,
the taxes laid on the collective and state farms
are always the most significant. Indeed, it happens
quite frequently that the cooperatives or state
farms take the leading role and build houses for
their employees outside the existing villages.
     During the course of the modernization of
agriculture, the infrastructural sensitivity of
agriculture itself had increased greatly. In my
opinion, in the most productive regions of the
Great Plain, the reason for the significant underutilization of the available soil capacity is the
underdeveloped infrastructure. Thus, for example,
the demand of agriculture for electric energy, gas,
telephones and telexes has grown enormously, just
as much as that of the villages. The raw products
released by agriculture increase by 3 to 5 percent
annually and, according to forecasts, are expected
to double by the turn of the century. Such growth
cannot be expected of any other mineral product.
The transportation demand of agricultural goods
takes place unevenly in time and place, but continuously and everywhere. The improvement in the rural
road network is thus demanded by the necessity of
collecting and transporting the product as well as
by the increased demands for human labor-transpor-
tation within the large modern agricultural units.
The modernization of agriculture, therefore, can
conceivably quicken the development of the rural
infrastructure and can significantly contribute to
the betterment of rural existence.

Development of Regional Ties
Within the Rural Sphere

The modernization of agriculture significantly increased the hitherto exceedingly thin network of
contact among the various villages. It is important
to note that the ties between city and village are

no longer restricted to a trade of industrial for
agricultural goods or the utilization of urban
services. The agro-industrial integration tied the
urban and rural settlements into a single economic
chain: the food produced in the villages, the
processing industries partly in the cities, and the
industries serving agriculture with machinery, are
all tied together in organized forms. Another
aspect of urban-rural ties is derived from the fact
that 15 to 20 percent of agricultural specialists
now reside in the cities and commute to nearby
villages.

Moreover, the organized and continuous ties
among the villages have also taken place within the
confines of modern agriculture. Naturally, the
closest ties exist among those villages that are
located under the umbrella of a common farm. Many
types of economic cooperation have developed among
the various neighboring villages; for example,
jointly owned slaughterhouses, silos, warehouses,
or processing plants have been created. The most
complex ties among the villages, however, have been
developed through the adaptation of various production systems.

The production systems are designed to produce
a single commodity or crop on a modern, industrial-
like basis and serve as the most important basis of
horizontal integration. The center of the system is
most often a state farm or an exceedingly good
cooperative farm. The innovational center, therefore, remains within agriculture. Between the center of the system and its member units, the ties
are continuous and well organized, since special
assistance, consulting, and control are also included as a part of the center's activity. These
production systems are not developed regionally;
partner members are tied to a single system from
vastly different parts of the state.

In 1977 sixty-seven production systems were
operated: of these, twenty-one were crop production, twenty-seven were fruit and viticultural, and
nineteen were animal-husbandry systems; 86.6 percent of the state and 78.3 percent of the cooperative farms were members of one or more system.
One-third of the arable plow land and one-fifth of
the fruit and vineyards were worked by these modern
systems. Even in animal husbandry, the modern
system's management approaches were extended to 50
percent of all animals produced on a large-scale
basis.

Summary

Rural settlement has been harmonious with the
surrounding agriculture. The size of the village,
the topographical location of its population, its
internal structure, its houses, the character of
its services and the road network were all adopted
to the needs of agriculture itself. This traditional harmony – precisely because of the slow development of agriculture and the stagnation or rural
society – survived even in the first-half of the
twentieth century. The socioeconomic development of
the last thirty years has broken that traditional
harmony once and for all. As a result of the rapid
social transformation, the rural settlements must
serve more than the needs of agriculture. The
demands of agriculture, however, have also been
altered significantly as a result of technological
development and agroindustrial integration during
the last decade. The technical infrastructural
development of the rural sphere was considerably
behind that of agriculture; in fact, it was not
even able to satisfy the needs of the latter. As a
result, tension developed between the village and
agriculture. The resolution of this tension is
necessary both because of the needs for increased
agricultural output and because of the need to
improve the standard of living of the rural population. The rural settlement network attempts to
adopt quite spontaneously to the changed social
environment for which it provides a framework. It
is high time to attempt to define the rural model
and the rural-urban relationship of the developed
socialist system, in order to allow our settlement
planners to plan for the entire settlement network,
not to restrict its development merely to our
cities and urban spheres.

REFERENCES

Andorka, R. "A kozsegi nepesseg tarsadalmi jellem-
zoi" (Social Characteristics of the Rural
Population). Tarsadalmi Szemle 29, nos. 8-9
(1974): 69-75.

Barta, Gy. "Hungary's Rural Industry." In Urban
Development in the USA and Hungary. Studies
in Geography in Hungary No. 14, (Budapest:
Akademiai Kiado, 1978): 293-301.

Barta, Gy. "Hungary's Rural Industry." In Urban
Development in the USA and Hungary. Studies
in Geography in Hungary No. 14, (Budapest:
Akademiai Kiado, 1978): 293-301.

Beluszky, P. "Falusi telepuleseink osztalyozasa"
(A Classification of Our Rural Settlements).
Foldrajzi Ertesito 14, no. 2 (1965): 149-163.

Enyedi, Gy. "A magyar falu atalakulasa" (The
Transformation of the Hungarian Village). Foldrajzi Kozlemenyek 23, no. 2 (1975): 109-124.

Enyedi, Gy. "A korszeru falufejlesztes elvei es
legujabb tendenciai" (Principles and New Tendencies of the Modern Rural Development).
Teruletrendezes 24, no. 4 (1977): 11-15.

Enyedi, Gy. "A mezogazdasag fejlodesenek hatasa a
falusi telepulesekre" (The Influences of the
Agricultural Development upon the Rural Settlements). Varosepites 18, no. 1 (1978): 2-4.

Lettrich, E. "Les traits characteristiques de
l'urbanisation en Hongrie." In Urbanization
in Europe. (Budapest: Akademiai Kiado, 1975):
133-137.

Sipos, A. "Az agrar-ipari integracio" (The Agro-
Industrial Integration). Kozgazdasagi Szemle
24, no. 6 (1977): 613-627.

18 The Development of Industry and the Industrialization of Villages: The Hungarian Example

Gyorgyi Barta

Industrialization accelerated the economic development of the village at an unprecedented rate and
had an enormous impact on the transformation of
rural society. Important technological changes in
Hungarian agriculture, as well as the introduction
of closed industrial production systems, proved to
be a sine qua non for creating and effectively
running large-scale socialist farms. Aside from
cultivation and production, these big agricultural
estates also have certain manufacturing and trading
activities; thus, a large proportion, about
one-third, of all those employed in agriculture
perform jobs which are distinctly industrial and
tertiary.

Auxiliary, nonagricultural activities were introduced in agricultural cooperatives both to
strengthen their economic position and to assure
full-time employment for the cooperative members.
There is a wide variation over regions and over
farms, but in sum, about 30 percent of all net
production of agricultural cooperatives derives
from such auxiliary activities. In our estimate,
150,000 to 180,000 cooperative members perform
industrial or service jobs either on their own or
within the framework of production cooperation
between industrial plants and the farmers' cooperative.

Within the spatial decentralization process of
industry, the location of manufacturing plants in
villages has become widespread. About 20 percent
of all state and cooperative manufacturing industry

at the beginning of the 1980s is located in villages. This proportion is approximately the same regarding rural employment and fixed assets.

The most characteristic feature of rural industry is its organizational dependence: most industries located in villages form branches of larger enterprises that have their head offices in towns. The industrial unit in a settlement, the plant, may be either independent (firm) or part of a larger enterprise (branch). This branch characteristic has been further strengthened in recent years. It was the branch characteristic that shaped the present structure of rural industry decisively and that had the greatest effect on the formation of the sectoral structure of rural industry. At the beginning of this century, the big industries in villages were mainly made of sectors localized near sources of raw materials (mining, metallurgy, and the building material industry), while today rural industry assumes the national sectoral pattern of industry, deviating very little from it. Since 1970, similar to changes in the national sectoral structure, the weight of mining dropped while that of heavy and light industries increased within the sectoral structure of rural industry.

To some extent, the branch character also influences the employment structure of rural industry. Vertical distribution of labor between the parent company and its branch is frequent, and it renders unneccessary the formation of a managerial or even a professional staff within the branch.

Changes in the spatial structure of rural industry are also increasingly influenced by branch character. The geographical situation of the industry-locating center affects the choice of settlements for locating branches. This explains why the greater part of rural industry is concentrated in areas where industry is relatively developed and why about 40 percent of rural industrial jobs can be found within the daily commuting rings of major towns.

The branch character makes its effect felt also in the development of rural industry. Financial soundness and the development prospects of the locating firms determine the size and the technical-equipment level of the branch. The independence enjoyed by rural industry is limited largely by external factors and it is almost

entirely exposed to the effects of economic fluctuations. Consequently, rural industry is remarkably
unstable; it is surprising how often industrial
plants are opened and closed in villages.

The qualitative features of rural industry,
that is, the conditions of industrial production,
can be attributed, first of all, to the branch
character and only secondly to the characteristics
of the rural labor force. The fact that industrial
workers in villages have lower qualifications than
those in towns expresses both the lack of demand
for qualified labor on the part of the firm and the
below-average education level of the rural population. Low qualification levels of workers and low
technical standards of equipment explain meager
productivity, which shows up sharply in the unfavorable wages of industrial workers in rural areas.
This chapter analyzes the relationships and causal-
ities between industry located in villages (rural
industrial plants with more than 10 workers) and
(a) the developments in the village and (b) changes
in the living conditions of rural population.

Industry exerts a direct influence when its
activity or products serve to satisfy local population needs. An indirect effect is made, not by the
industrial activity as such, but by its capacity to
create jobs, for example, to generate income and
contribute to the physical developments of the
village.

Before World War II, small-scale industry was
extremely significant in Hungary. Artisans in villages played the most important role in supplying
the rural population with industrial goods and in
providing industrial services. The number of artisans in each village varied by the population size
and geographical situation of the village, but one
could rarely find a village with no artisans at
all. Rural population in Hungary could almost
everywhere obtain local industrial goods and services, although they were not of the highest
standard.

Following 1945, drastic restrictions imposed
on small-scale industry through such measures as
the introduction of a prohibitive tax created a gap
in the supply of rural population, which has not
been filled since. Today, the number of artisans
in villages is close to eighteen thousand persons,
that is, three artisans per 1,000 rural inhabitants. Those trades which were the most widespread
before World War II, such as locksmith, carpenter,

shoemaker, bootmaker, and tailor, have preserved their importance. New trades born by modern technology, such as automobile servicing and radio and TV repair, have also been created. Naturally, numerous trades are disappearing, such as potter, hand spinner and weaver, and miller.

Following World War II, the role of artisans was only partly taken over by the industrial servicing cooperatives (for example, cooperatives for repairing and servicing household appliances, TVs, radios, and automobiles), which were mostly formed by groupings of former artisans. Such cooperatives, however, exist only in larger settlements today, and thus a long series of villages is virtually compelled to miss those services entirely or can only enjoy them at a high cost and through great difficulties.

The auxiliary activities of industrial character in agricultural cooperatives are not yet of great importance, but are developing vigorously and offer more and more room for supplying local inhabitants as well (for example, baking, meat and other food-processing, industrial plants selling part of their products on the spot, and the construction industry).

Finally, however, industrial goods purchased by the rural population are almost entirely produced by big industries, and industrial services are usually only available in towns and larger villages. Today it is not merely the kind of industrial goods and services, but their composition and quality that separate the consumption of the rural population from that of urban dwellers.

Industry generated by local needs is not significant in villages, considering either the size of industrial activity or its share in rural consumption. The indirect effect of industry located in villages is far more important. Precisely due to its indirect character, however, it is not easy to assess and evaluate the effect of rural industrial activity; similar effects can be felt by such factors as the other economic sectors present in the settlement, the villages' geographical situation, and its location with regard to transport facilities.

(A) The effect of industry located in villages on their overall development.

Industrialization in the villages brought about

various changes in rural living conditions. It created new jobs and increased their variety. Industrial work usually provided better working conditions than agricultural work (such as regulated,
shorter working hours, often cleaner work, and less
physical stress, which is particularly important
for women). Last but not least, it increased the
family's income.

From among the many beneficial effects, we
shall concentrate on demonstrating the role which
industry has played in increasing the level of
employment and in enlarging personal income resources in rural areas.

Following World War II, the employment level,
the number of earners, increased substantially in
rural areas. Nevertheless, due to the still existing economic differences between the city and the
village, to the peculiarities of rural existence,
and also to the disputable statistical recording,
the employment level of the rural population is
lower than that of urban dwellers and much lower
than that of the inhabitants of Budapest (see Table
18.1).

Table 18.1.   The Proportion of Wage-Earners
              by Types of Settlement (percent)

Year	Budapest	Towns	Villages
1949	53.6	45.7	45.8
1960	63.5	52.4	48.6
1970	73.9	62.5	57.4

Expanding female employment has been characteristic
of the increasing level of employment in general,
as well as within rural areas (see Table 18.2).

Table 18.2.   Economic Activity of Women
              1941-1070

Proportion of Female Earners
Within Total Female Population (percent)

Year	National Average	Villages
1941	26.0	22.2
1960	37.0	31.2
1970	53.1	44.5

To assess the role and importance of industrial advances, the effect of industrialization in villages should be compared to other factors that influenced the increase of employment level in several directions.

The most significant change was a sudden increase in the number of inactive earners, out of which the highest proportion falls to pensioners. The increase in the number and in the proportion of those retired depends on the strength of the national economy. Their number and proportion may increase by the aging of population (in Hungary the retiring age is fiftyfor women and sixty for men), and by enlarging eligibility. In rural areas, the latter factor played a particularly important role in the growth of number of the inactive earners. Retirement was extended to members of agricultural cooperatives in 1966, and that resulted in a slight decrease in the rate of activity (from 46.2 percent in 1960 to 45.2 percent in 1970) and a five-fold increase in the rate of inactive earners (from 2.4 percent to 12.0 percent). All in all, between 1960 and 1970, more than 90 percent of the total increase in the number of earners resulted from widening the circle of inactive earners.

The growing number of commuters had an effect of somewhat smaller significance than the growth of the number of inactive workers, though still significant from the perspective or rural industrialization. Since the end of the fifties, commuting has involved large numbers of people in Hungary, with both the absolute and the relative number of commuters growing continuously. In 1970 the number of commuters reached almost one million, and their proportion within the active-earners group increased from 12.5 percent in 1960 to 20.0 percent by 1970. Obviously, a large proportion, about 90 percent, of commuters reside in villages. The balance of in- and out-commuters to and from villages shows an ever-growing out-commuting surplus; from 1960 to 1970 the out-commuting surplus of villages increased by more than 70 percent to more than 200,000 persons. Commuting is particularly widespread in the industrial sector. In 1970, more than 60 percent of out-commuters of villages found jobs in industry. Industry located in villages, however, also employs commuters in a growing number. In 1970 about 40 percent of industrial jobs in the villages was occupied by commuters.

Another factor, which does not reflect the economic strengthening of the village but has contributed to the increase of employment level, is a drop in the absolute number of rural population since the mid-sixties. Between 1960 and 1970, the rural population decreased by almost 3 percent; hence, there is no more rural overpopulation anywhere in the country.

Economic development in rural areas resulted, first of all, in a structural change of rural economy (see Table 18.3), because the decreasing need for labor in agriculture was only partially counterbalanced by developments in other economic sectors.

Table 18.3.   The Structure of Economy in Villages
              (1960-1970)*   (percent)

Sector	1960	1970
Agriculture	69.2	41.5
Industry and Construction	19.7	22.4
Others	11.1	36.1
Total	100.0	100.0

*Based on the number of people employed in the economy of villages

The above structural change resulted both from an absolute decrease in the number of active earners (agricultural employment dropped by half a million) and from a 10 percent increase in the number of jobs in the nonagricultural sectors. Forty-one percent of new jobs were created in industry and a larger part in the tertiary sector.

It is not possible to discern the effect of the augmentation of jobs in villages on the level and structure of rural families' incomes and to discover whether industrial earnings only supplement agricultural incomes. Nevertheless, available data demonstrate that in 1975 the sum of salaries paid off in industries located in villages (more than nine thousand million forints for 360,000 industrial workers) ran to almost 60 percent of the income earned in the agricultural cooperatives of villages by more than half a million people. Rural earners have certainly other means as well to supplement their income (such as the household farm, income from the tertiary sector, commuters' income, and pensions). Industry located in villages contributes clearly to improving and stabilizing the financial situation of rural families. This is true even if the average income level in rural industry falls short of the urban or the national level. This finds expression in the fact that, while more than 20 percent of industrial workers have their jobs in villages, their share in the total industrial salaries is hardly over 17 percent.

The effect of the presence of industry cannot be always directly measured by changes in the rural inhabitants' living conditions or by their increasing living standards. It is also worthwhile to examine what industrial location "brings" to the village, whether it helps its economic development, and if the local administration is interested in attracting industry to the village. Our analysis here is directed toward two groups of questions: (a) Does the village benefit from the output of local industry? More precisely, what kind of taxes or charges does the industry have to pay to the local council? (b) Do the appearance, activity, and taxation of industry improve infrastructure in the village?

There are three payments within the taxation system of industries which have spatial aspect, that is, which are paid by industry to the "spatial" council, namely:

1. Takeover charge and rent for the use of land;

2. A part of the depreciation of fixed assets at enterprises controlled by councils (40 percent of amortization);

3. One-third of contributions for the development of towns and villages levied on industrial firms and industrial cooperatives located within the jurisdiction of councils.

According to the financial rules of county councils, only the first type of payment may remain with the council where the industrial plant is located, but that possibility is seldom realized: on the one hand, most county councils concentrate on this type of payment; on the other hand, most industrial plants located in villages constitute branches of others; thus, the takeover charge and on this type of payment; on the other hand, most industrial plants located in villages constitute branches of others; thus, the takeover charge and rent for land use is paid by their centers to the council where the headquarters are located.

The other two types of taxation money are collected by county councils all over the country. Their spatial redistribution follows settlement hierarchy, which always works to the disadvantage of villages ranking at the bottom of the hierarchical ladder. Consequently, villages do not benefit from the industrial plant as a taxable income resource. In this respect, rural industrialization favors mostly the development of towns, especially the larger urban centers.

Besides obligatory payments, industrial plants often assign larger sums out of their own development funds to councils or to societal organizations. This money is used mainly for settlement development or for the development of cultural institutions. It may happen that the settlement where the donor industrial plant is located also benefits from the mentioned resources,. but such "generosity" on the part of industrial firms is incidental; it is not obligatory, and the sum offered is not too significant. It can, however, result in a slight improvement of the village's neglected situation that follows from the centralized system of spatial development.

The infrastructural supply level of villages with industries is higher than that of villages on the average. It was attempted to find a relationship determining whether the presence of industry affected directly or indirectly the development of infrastructure or the relatively higher supply standards derived from other factors. It is, in fact, the already existing infrastructure that attracted industries to settle in the village.

When analyzing infrastructure as related to industry and industrialization, a distinction should be made between technical infrastructure, indispensable for industrial activity, and infrastructure for population use, which might have been induced by industry.

It follows from the taxation system of industrial firms, which has been briefly outlined above, that the presence of industry can hardly or not at all contribute to the improvement of infrastructure for population use in the village.

As far as technical infrastructural investments are concerned, they are mostly financed from the development funds of the industrial firms. From the objectives of rural industrialization (search for additional manpower at minimal capital expenditures) as well as from the character of industries located in villages (most of them are branches depending on industrial firms residing in urban areas, and the majority consist of small plants with 10 to 200 workers), it follows that the proportion of immovable properties within the total value of fixed assets, namely, that part which contains elements of infrastructure, is smaller in the villages.

For 280 industrial plants established between 1972 and 1975, it is possible to show that the proportion of immovable properties (that is, machinery deducted from the total value of fixed assets) increases with the size of plants (figure 18.1).

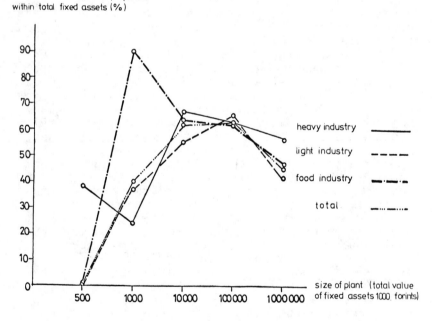

Figure 18.1.   Village Industry: Size and Immovable Properties

Finally, the conclusion can be drawn that opening industrial plants directly did not improve, or improved only to a very small extent, the development of infrastructure in villages.

This statement is supported by a secondary analysis concerning the infrastructure of villages during the period preceding industrialization. Selected elements of infrastructure in 450 industrializing villages (for example, where new industrial plants had been opened between 1970 and 1975) were compared with average standards of all Hungarian villages (see Table 18.4). It can be seen from Table 18.4 that industrializing villages had already reached and mostly exceeded the average level of all villages before their new industrial plants were opened.

Industrialization does not really explain the process of differentiation among villages concerning their infrastructural supply. A proper answer can only be given if all other resources of infrastructural investments are explored. The important resources of infrastructural investments in villages are the following:

1.   Population taxes;

2.   Taxes and other payments of economic units;

3.   Subsidies given by the county council.

The second and third resources cannot be considered general (only those villages benefit from them which give room for the headquarters of an economic unit); they are incidental and they are very small in most villages. In effect, it is the population taxes (contribution to local development, land tax, tax on road and public utility development) that village councils dispose of completely and in all villages. And the amount of population taxes depends on the number of inhabitants.

The number of inhabitants in industrializing villages is usually larger than the average number of inhabitants in Hungarian villages. In 1970, while industrializing villages had an average population size of about 4,000 persons, the national average in villages amounted to 1,700 inhabitants.

This fact explains why infrastructural supply standards are higher in industrializing villages.

Table 18.4.   Infrastructure Supply Standards of Industrializing Villages
              Preceding the Introduction of New Plants, 1970.

Elements of Infrastructure	Industrializing Villages (1)	National Average of Villages (2)	(1) as percent of National Average (2)
Proportion of dwellings supplied by:			
– electricity (4)	85.9	86.2	99.7
– water conduit (%)	17.7	10.8	163.9
– sewage (%)	14.2	11.8	120.3
– gas	40.9	40.5	101.0
No. of dwellings per 100 inhabitants	31.1	31.3	99.4
Proportion of state owned buildings	9.1	10.0	91.0
Electricity consumption per 1,000 inhabitants (MWh)	87.0	86.6	100.5
No. of places in creches and kinder-gartens per 100 children aged 0-6 years	28.4	25.0	113.6
No. of children aged 6-14 years per one classroom in elementary schools	32.8	40.7	80.5
No. of consultation hours by special-ists per 1,000 inhabitants	0.8	1.19	67.2

Sources: Census data, 1960 and 1970;
           Teruleti Statisztikai Zsebkonyv 1971 /Regional Statistical Pocketbook 1971/
           (Budapest: KSH, 1971);
           Helyzetkep az orszag varosairol /General Survey of Hungarian Towns/, Budapest:
           KSH, 1971)

Positive Effects of Rural Industrialization
On the National Socioeconomic Development

The changes in the spatial allocation of productive
forces, speeded up by industrialization from the
end of the nineteenth century and by the new
socioeconomic processes following World War II,
were accompanied by the spatial redistribution of
population. This process was manifest in the relative decrease of rural population (its proportion
of the total population), which has been going on
constantly but at a changing rate. At the turn of
the century, 66.4 percent, and in 1949, 61.6
percent, of the Hungarian population lived in
villages. This proportion dropped to 48.0 percent
by 1977.

Since the mid-sixties, the number of rural
residents has been falling in absolute terms as
well; indeed, between 1960 and 1976, it fell by
almost 300,000 persons. The absolute and relative
population decrease resulted mainly from a powerful
out-migration. Between 1949 and 1960, 8.4 percent,
and between 1960 and 1970, almost 10 percent of
rural population moved to towns. After 1970 this
process continued, though at a lower rate; between
1970 and 1976 more than 4 percent of the rural
population migrated to urban areas (see Table
18.5).

Table 18.5. Spatial Variations in Natural Increase
(1949-1975)

Percent Annual Natural Change

Periods	Villages	Towns without Budapest	Budapest
1949-1960	1.02	0.92	0.48
1960-1970	0.45	0.47	-0.10
1970-1975	0.44	0.75	0.08

Sources: Census data 1960 and 1970;

Az allando es lakonepesseg szama, a termeszetes
nepmozgalom fobb adatai kozsegenkent 1970-1976
(Permanent and resident population, vital statistics by villages 1970-1976) (Budapest: KSH, 1977).

The rapidly falling natural change of rural
population could not counter-balance these migration losses. The number of village inhabitants
considerably decreased due to the fact that the
number of villages also dropped. Between 1900 and
1977, fifty-two villages had become reclassified as
towns (accounting for more than half of the total
number of the present Hungarian towns). The total
population of these fifty-two villages measured at
the time of their reclassification amounts to one
million persons.

The settlement network characteristics, the
geographic and transport situations of villages,
the economic development of villages and of their
region, and particularly the population sizes of
villages resulted in demographic changes showing
wide variety by regions and by villages (see Table
18.6).

Table 18.6   Changes in the Resident Population of
             Villages, 1960-1970.
                      (percent)

| Population size | Growth | | Net |
groups (persons)	Actual	Natural	Migration
0 - 499	-16.4	1.9	-18.3
500 - 999	-11.7	3.2	-14.9
1,000 - 1,499	- 9.8	4.2	-13.9
1,500 - 1,999	- 8.1	4.4	-12.5
2,000 - 2,999	- 6.8	4.9	-11.7
3,000 - 4,999	- 5.4	4.7	-10.1
5,000 - 9,999	- 0.4	5.2	- 5.6
10,000	+ 6.3	5.1	+ 1.2
Average	- 5.4	4.5	- 9.9

Source: Census data, 1970

Let us now examine the demographic "behavior" of
industrializing villages in order to discern
whether the demographic processes of these villages
deviate from the average of all villages. There is
no doubt that if some kind of positive deviation -
the expected signs of demographic stability - could
be demonstrated, it could not be directly linked to
their industrialization.

Industrializing villages (somewhat more than a
thousand villages can be classified as such) in-

volves two-thirds of all villages and 60 percent of
the rural population. Between 1960 and 1970, the
size of the population living in these one thousand
villages did not decrease, but stagnated; their
population increased by merely 0.5 percent, that
is, by 16,000 persons, between 1960 and 1970,
disregarding those villages which became towns in
the meantime.

There are, of course, differences between larger and smaller villages as far as population
changes of industrializing villages are concerned,
but those differences are not so sharp, and losses
are much smaller in size groups under 3,000 inhabitants than in similar size groups of all villages
in Hungary (see Table 18.7).

Table 18.7. Changes in the Population of Industrializing Villages
                      1960-1970. (percent)

Population Size Groups (Persons)	Distribution of Population, 1970 (%)		Shift In Share Within Industrializing Villages (%)[+]
	Industrializing Villages	All Villages	
- 499	0.4	4.1	+3.8
500 - 1,000	2.3	10.8	-7.4
1,000 - 1,500	4.2	10.7	-1.6
1,500 - 2,000	5.7	10.3	-2.5
2,000 - 3,000	12.0	14.8	-2.1
3,000 - 5,000	23.2	17.6	-0.9
5,000 - 10,000	31.7	18.7	-0.4
10,000 -	20.5	13.0	+3.9
	100.0	100.0	+0.5

[*]for 1960 administrative areas

At the same time, industrializing villages are characterized not only by a stable demographic situation, but also by a rather marked change in the tendency of their demographic processes. Having compared the periods of 1960-1970 and 1970-1976 with regard to the direction and pace of population changes, five groups of villages can be delineated (Table 18.8).

Table 18.8.   Direction and Pace of Population
              Changes in Villages.

	Distribution by the Number of Villages
Increasing pace of growth	27.0%
Similar pace of growth	8.2%
Stagnating	15.5%
Similar pace of loss	33.8%
Increasing pace of loss	15.5%
	100.0%

Population dynamics of the first and the fifth groups - 42.5 percent of industrializing villages - changed in the first half of the seventies as compared to earlier years; 27 percent of villages showed an acceleration of population growth and 15.5 percent showed an acceleration of population loss.

The rural settlement network is also changing its structure, since dynamically developing villages form new growth nuclei. In 1970, for example, 23 percent of industrial villages - more than two hundred villages - developed into local centers; their industries also employ commuters from neighboring villages. These rural centers of attraction are located in areas where urban centers are scarce and where the population size of villages is very small. Industry is one of the most efficient settlement-developing factors. For example, in the 23 villages which were classified as towns during the past ten years, the number of locally employed industrial workers reached 54,000 persons, 39 percent of the total employment. Industry, with few exceptions .(for example, mining villages), constitutes an additional or a major function within the economy of a village, but it is usually not an exclusive activity.

With the spread of industrialization, the number of multifunctional rural settlements providing possibilities to maintain and preserve the vitality of communities increases. In an important proportion of villages, consequently, industry has had a very positive impact on the living conditions of the population. It gives a stimulus to the economic development of the village, which contributes to eliminating its underdevelopment and diminishing the differences between the town and the village.

REFERENCES

Andorka, R., A magyar kozsegek tarsadalmanak atalakulasa (Societal Transformation of Hungarian Villages), (Budapest: Magveto Kiado, 1979), p. 166.

Arvay, J. (ed.), A magyar ipar (The Hungarian Industry) (Budapest: N.P., 1976).

Barta, Gy., "A falusi ipar teruleti problemai" (Spatial Problems of Rural Industry), Teruleti Statisztika no. 4 (1979): 377-391.

Barta, Gy., "A falusi ipar altalanos helyzete es teruleti problemai" (Overall Situation and Spatial Problems of Rural Industry) (Manuscript, FKI, Budapest, 1979): 146.

Barta, Gy., "A falusi ipari kutatas fontossaga es idoszerusege" (Importance and Relevance of the Research of Rural Industry), Teruleti Kutatasok no. 2 (1979): 35-52.

Berend, T.I. and Gy. Ranki, A magyar gazdasag szaz eve (One Hundred Years of Hungarian Economy), (Budapest: Kossuth-KJK, 1972).

Enyedi, Gy., "A magyar falu atalakulasa" (Transformation of the Hungarian Village), Foldrajzi Kozlemenyek no. 2 (1975): 109-125.

Enyedi,   Gy.,   "Falukutatasok   a   falufejlesztesert"
(Village   Research   for   Village   Development),.
Foldrajzi Kozlemenyek   nos.   3-4   (1975):
269-277.

Enyedi, Gy., "Rural Transformation in Hungary." In
Rural Transformation in Hungary and Poland   (A
Polish-Hungarian Seminar in 1978) (Budapest:
Geographical Research Institute, 1979): 4-12.

Kulcsar, V. (ed.), A valtozo falu (The Changing
Village) (Budapest: Gondolat, 1976): 341.

Meszaros, R., "Ipari tevekenyseg Bacs-Kiskun megye
mezogazdasagi termeloszovetkezeteiben" (Industrial Activity in the Agricultural Cooperatives of County Bacs-Kiskun). (Manuscript,
FKI, Budapest, 1979): 26.

Ranki, Gy., "A kisipar szerepe a magyar kapitalista
fejlodesben" (The Role of Small-Scale Industry
in   Hungarian   Capitalist   Development).,
Tortenei Szemle 7, no. 2 (1964): 423-451.

Szego, A., Teruleti erdek, teruleti igazgatas es
fejlesztesi politika (Spatial Interest, Spatial Administration and Development Policy)
(Budapest: Tanacsigazgatasi Szervezeti   Inte-
zet, 1974).

Vagi, G., "'Mit er' egy kozseg? 'Mit er' egy
megye? - A tanacsi fejlesztesek intezmenyi
szerkezeterol" (What is a Village "Worth"?
What is a County "Worth"? - On the institutional Structure of Council Developments),
Kozgazdasagi Szemle nos. 7-8 (1975): 882-899.

19 The Industrialization of Agriculture in the Kibbutzim: The Case of Israel

Avinoam Meir

Kibbutzim, Israel's communal rural settlements, are frequently thought to employ agriculture as their main economic venture. This image is the result of the distinctive type of rural life in these settlements and the persistence to the present day of their successful agricultural endeavors. However, in a process that began in the early 1930s and accelerated in the 1960s, kibbutzim have been adopting industry as a supplementary and, on occasion, even primary economic activity. The purpose of this chapter is to inspect and analyze kibbutz industry through a discussion of its origins, growth, and nature, and the general factors that have brought about the adoption of this economic diversification.

THE UNIQUENESS OF KIBBUTZ INDUSTRY

The uniqueness of kibbutz industry derives from its being a social and economic rural cooperative. The kibbutz's cooperative nature is responsible for its success as a regional development agent in Israel. The kibbutz has always served as a pioneer force in settling frontier areas, a goal of top political priority for the Israeli government. Were it not for the group-action characteristic of such settlement efforts, Israel's sovereignty over certain areas might well have been seriously doubted at the time of the creation of the state.

It is this group-action character which is

259

unique to kibbutz industry. It was established with
the welfare of the social group being the primary
goal. The kibbutz's purely socialistic version of
industry, characterized by group ownership, management, and supervision, was encouraged by the so-
cialist-oriented Labour government. Since capital
resources were so chronically lacking in these
frontier areas, the communal pool of manpower
provided at least a partial substitute for those
resources necessary for industrialization. The performance of kibbutzim as agricultural entrepreneurs
established institutional faith within the government in their capability to make the change toward
an agro-industrial society within the framework of
rural life. Consequently, the Israeli government
has been financially supporting industry adoption
by kibbutzim since the mid-1950s, utilizing them
within its program to disperse industry to periph-
eral rural areas.

ORIGINS AND GROWTH OF RURAL KIBBUTZ INDUSTRY

Although agriculture is still the primary economic
activity in many kibbutzim, the idea of industry as
a possible economic venture has always been well
rooted in ideological manifestos of the kibbutz
movement. (1) The existing meager evidence points
to the late 1920s and early 1930s as the period
when embryonic stages of the process began.
    The half-century long industrialization-dif-
fusion process can be divided into four periods.
The first period took place in the 1930s, when
manufacturing shops and small-scale factories were
established to supply local and internal Kibbutz
needs and limited marketing. Most of the
small-scale factories were engaged in processing
agricultural produce and manufacturing basic agricultural equipment. (2) The second period took
place in the 1940s. During World War II, a resource
mobilization was ordered by the British Mandate
which, along with expansion of existing shops, gave
rise to the establishment of new food, textile,
metal, and packing material plants. At the same
time, the struggle for independence from the British Mandate created the need to establish illegal
underground munitions shops, which were transformed
into legal metal factories subsequent to the establishment of the State of Israel in 1948. (3)

The third period of industrialization began after 1948 and continued during the 1950s. This period witnessed an enormous population increase in Israel, due to massive immigration. The growing market, along with a shortage of foreign currency, forced the Israeli government to impose an import substitution program, to which kibbutzim began to intensively respond only in the mid-1950s, (4) and to establish assembly plants with imported technology.

The final period began in the late 1950s. Domestic technological and managerial knowledge, as well as information regarding the success of earlier industry experience in kibbutzim, were transmitted among other kibbutzim. A significant impetus for the dissemination of such information was provided by the establishment of the Kibbutz Industry Association as the consulting agency in 1963. Since that year, the process of industrialization adoption by kibbutzim has accelerated in the number of both plants and employees. (5)

Thus, there have been two types of industrialization adoption among kibbutzim. The first is an adoption of industry by first establishing a small workshop as an auxiliary service; this manner of adoption was characteristic of the early periods of the process. The second type of adoption involved the establishment of a factory with market orientation being the primary goal; this has been more common in the latter periods of the process, especially since the early 1960s.

Figure 19.1 presents the temporal trend in the adoption of industry by kibbutzim. The trend in the percentage of total productive labor force allocated to manufacturing and industry reveals a highly variable increase rate. The most significant increase occurred during the 1960s, although the mid-1960s economic recession and the 1973 war severely affected the growth trend. By 1978, about 38 percent of all kibbutz productive labor was allocated to industry. On the other hand, using the presence of at least one factory in a kibbutz as a criterion for the adoption of the innovation (Figure 19.1), it seems that the process of diffusion of industrialization among kibbutzim is approaching its end. Recent adoption growth rates indicate that the 1980s will witness an integrated industrial-agricultural economic system common to all kibbutzim.

Table 19.1 Kibbutz Factories by Industry, 1947-1975 (in percent).

Industry	1947	1955	1965	1969	1972	1975
Metal and Printing	36.5	43.3	36.5	31.0	29.3	29.8
Electronics	-	0.3	5.7	9.1	11.6	11.0
Timber and Furniture	34.1	29.7	12.1	8.5	6.8	7.0
Plastic and Rubber	-	-	10.1	15.2	17.6	21.7
Food	11.1	9.3	15.2	11.5	9.0	7.7
Chemicals	2.4	1.4	2.5	3.0	2.5	1.4
Textile and Leather	7.9	4.2	7.0	5.4	5.1	5.1
Building Materials	2.4	4.3	8.2	4.8	5.1	4.0
Ornaments	-	-	-	-	-	4.0
Misc.	5.6	7.1	3.1	10.9	9.4	7.7

Source: 1947 – Y. Don, Industrialization in Advanced Rural
Communities (paper presented at Fourth World
Congress of Rural Sociology, Turon, Poland,
1976).

1955 – Compiled from unpublished data of Audit Union of
the Woof the Worker's Agricultural Societies,
Ltd.

1965 – Compiled from data in A. Shachal, "Kibbutz
Industry: Development, Structure, and Problems"
(unpublished paper at University of Tel Aviv,
1968)

1969 – 75 – Compiled from Kibbutz Industry Association,
Ltd., Annual Reports, 1968-1976 (Tel Aviv).

THE NATURE OF KIBBUTZ INDUSTRY

The nature of kibbutz industry has undergone considerable changes throughout the adoption process
and refers to three aspects of this innovation:
type of industry, size of production unit, and
market orientation.

Table 19.1 presents the trends in the industrial branches from the mid-1940s to the mid-1970s.
Several trends are apparent. Until the mid-1950s
kibbutz industry was strongly oriented toward metal
and furniture products, with a secondary emphasis
on food products, which was a residue of the
stronger emphasis that prevailed in the 1930s.
However, the metal and furniture industries experienced a strong relative decline in later periods.
The largest decline in the relative share was
experienced by the furniture industry and, to a
lesser extent, the metal and food industries. On
the other hand, a considerable increase was experienced by electronics and plastic, and rubber industries. The rest of the industries remained quite
stable, reflecting a permanent market demand for
unsubstitutable products. Considerable attention
has been given in recent years to the planning and
developing of the optical industry by new adopters.

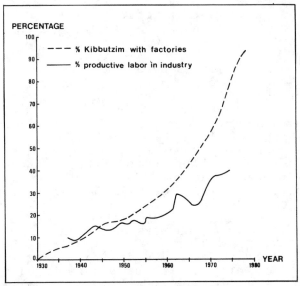

FIGURE 1 : THE ADOPTION OF INDUSTRIALIZATION BY KIBBUTZIM : 1930-1978.

Source: Compiled from unpublished data of Kibbutz Industry Association and Audit
Union of the Workers' Agricultural Societies, Ltd.

The size of the industrial production unit has
always been a matter of concern for kibbutz economic institutions for several reasons. Firstly, the
kibbutz as an economic production unit adopted
industrialization within the context of a constrained labor force that had to be allocated among
various activities. Secondly, financial resources
were, and continue to be, constrained despite the
accumulation of some physical and monetary capital
in kibbutzim since the late 1950s. (6) Thirdly,
although some kibbutzim have accumulated technical
and managerial skill, these were primarily agriculture oriented; furthermore, most kibbutzim have had
only a short skill-accumulation period. (7) Therefore, adopting the innovation made it compulsory to
keep plant size large enough to survive competition, but small enough to avoid economic growth at
the expense of other economic activities in the
kibbutz or of the social well-being of its members.
Such a goal could not have been achieved
without thorough planning. However, modern adoption
planning started only in the early 1960s. Prior to
that period, kibbutzim adopted industry in a hap-
hazard fashion (8) and plant size seldom reflected
labor capability. The result was the frequent
necessity to resort to external labor resources,
thus creating an ideologically delicate and controversial issue in kibbutz life.
Modern industry adoption planning began largely
as a result of the establishment of the Kibbutz
Industry Association in 1963. This inter-kibbutz
consulting firm has assumed a central role in the
decision-making process regarding the selection and
nature of the industry to be adopted. It has had an
indirect impact on the adoption process as a whole
by disseminating information and facilitating the
establishment of factories that would have otherwise been delayed. However, it must be noted that
the decision of a kibbutz to shift to industry is
reached independently of this consulting firm. The
timing of this decision depends largely on internal
trends within the particular kibbutz.

One of the consequences of planning the industrial activity (whether new or expanding) has been
a recent substantial reduction in plant size, as
indicated by Table 19.2. The available data indicate the probable continuation of a trend that
might have started in the early 1960s. The percentage of small factories (under thirty workers)

Table 19.2. Percentage Kibbutz Factories by Size (Number of Workers) 1969-1975.

Year	Size Under 10	10-30	30-50	50-100	100+
1969	14.0	33.0	28.0	14.0	11.0
1970	25.0	29.0	23.0	14.0	9.0
1971	18.0	31.0	26.0	17.0	8.0
1972	21.0	34.0	20.0	17.0	8.0
1973	24.0	32.0	20.0	17.0	7.0
1974	27.0	32.0	20.0	14.0	7.0
1975	24.0	36.0	18.0	14.0	7.0

Source: Kibbutz Industry Association, Ltd.

increased, while that of the larger size (over thirty workers) declined. The smallest factories (under ten workers) were ranked fourth in frequency in 1969 and rose to the second rank in 1975. The largest plants (over one hundred workers) remained the smallest group throughout, further declining in relative terms in 1975, indicating recent, relatively small labor-scale adoptions.

A further outcome of introducing modern planning techniques was a change in the marketing orientation of kibbutz industry. Some of these changes were already indicated by Table 19.1. The major substitution trends have brought about a refocusing in marketing efforts from domestic to foreign markets. The growing specialization of kibbutz factories and the rather limited size of the Israeli market, paved the road toward penetration of overseas markets. Unfortunately, no export data are available for the 1950s and 1960s, but between 1969 and 1975 kibbutz industry export growth rate was 20 percent higher than the national rate of 73 percent and 61 percent, respectively), while the total sales growth rate was 80 percent higher than the respective rate at the national level (90 percent and 50 percent, respectively). The emphasis on foreign markets is further demonstrated by the fact that by 1974, kibbutz industry was producing about 38 percent of the national plastic and rubber products (in monetary terms), but exported 62 percent of the national output; in timber and furniture, the respective figures were 23 and 52 percent. (9) Kibbutz industry, thus, has been assuming a dominant position in foreign markets among Israeli manufacturers in these industries.

The situation in the future, as judged by recent investment trends, will further favor the kibbutz industry. Investment growth rate in kibbutz industry between 1969 and 1975 was 122 percent, compared to a 22 percent increase at the national level. With the highly negative national balance of payments and relatively limited foreign currency reserves, the importance of this rural-type industry in the future will increase further.

DETERMINANTS OF INDUSTRY ADOPTION BY KIBBUTZIM

The factors that have been operating to cause the temporally differential adoption of this economic change, in addition to the obvious factor of the temporally differential establishment of kibbutz settlements, can be broadly divided into four groups: ideological, physio-ecological, economic, and demographic factors.

Ideological Factors

The fundamental principle of the kibbutz, which has remained largely unchanged, is that the kibbutz is a village based on agriculture, manufacturing, industry, and service occupations which are performed in the framework of cooperative and communal life. (10) From this it follows that a kibbutz was ideally supposed to become a relatively closed economic system, where as many individual and communal needs as possible were to be provided internally. This stated principle of self-labor has put the kibbutz economy under certain constraints. Strict adherence to this principle would have meant size stagnation, unfavorable conditions for economic expansion, and consequently, an increasing internal economic and social stress. In anticipation of such a plight, ideological variations began to form regarding this issue and, together with other ideological issues, brought about the emergence of several kibbutz federations, each having its own ideological principles.

One of the characteristics of kibbutzim affiliated with the largest federation, the Union of Kibbutzim, was the desire to keep the kibbutz as small and cohesive as possible, with agriculture the main and preferably the only economic activity. These kibbutzim did not favor fast growth; consequently, prospects for a balanced economy were slim in light of manpower constraints. A similar ideology was shared also by the smallest federation, that of the Religious Kibbutz. (11) At the other ideological extreme were kibbutzim affiliated with the United Kibbutz Federation. These were dedicated to the concept of a large kibbutz with unrestricted growth and development, encouraging maximum autarchy and minimum dependence on external resources. (12) Between these two extremes were kibbutzim affiliated with the National Kibbutz Federation. These kibbutzim favored a homogeneous society with

totally controlled economic growth, and, by nature,
favored controlled economic mix within the kibbutz.
(13) The outcome of these ideological variations
was a differential attitude toward industry adoption. Thus it was a kibbutz affiliated with the
United Kibbutz Federation, which favored unrestricted growth and development, that first incorporated an industry.

However, the role of ideological attitudes in
the adoption process soon lost its significance. By
the late 1930s adoption data no longer demonstrated
the expected differential adoption rates. Several
developments made it impossible for Kibbutzim to
maintain strict principles and also to survive in
the long run as a viable social and economic unit.
Consequently, ideological variations yielded to
other, more important factors in the process of
adopting industry.

Physio-Ecological Factors

In order to understand the role of the physical
environment in influencing the adoption of industry
by kibbutzim, it is necessary to refer briefly to
the spatial process of the establishment of new
kibbutzim, because to some extent the adoption
process took place simultaneously with the land-oc-
cupation process. The latter process, which began
in 1909 and ended in the mid-1960s (referring only
to pre-1967 borders), was characterized by a spatial shift that undertook the general north-south
direction. In general, areas in the northern and
north-central parts of the country were inhabited
first, and the emphasis on settlement efforts
shifted with time towards the southern part of the
country. (14)

Political considerations were the primary de-
terminants that guided the sequence of land occupation. Consequently, it was often necessary to
totally ignore ecological considerations in the
decision to locate new rural settlements. Due to
the available technology in agricultural and water
extraction in the 1920s and 1930s, the spatial
distribution of potentially favorable physical environments was highly variable, with areas in the
northern and north-central parts of the country
offering better opportunities than the south. The
amount of annual precipitation in Israel decreases
with distance from the Mediterranean Sea and in a
north-south direction. The semi-arid areas extend

to the north as far as the $31^{\circ}30'$ parallel. (15) Regarding soils, the best potential was offered by areas in the northern valleys of Izre'al and Hula and the Coastal Plain (see Figure 19.2). Some of these areas were swampy, and only draining would provide additional land for new settlements. (16) Land and water resources, however, were and are today severely constrained relative to the political need for land occupation.

Early kibbutzim were located in the Izre'al valley and the Coastal Plain, where both fertile soil and water supplies were abundant. These are the lowland areas of the subdistricts of Izre'el, Zefat, Kinneret, Haifa, and Hadera. The decision of the settlement authorities in the second half of the 1930s to begin a massive political effort of inhabiting the hilly and mountainous areas was reached in spite of unfavorable ecological conditions there. Furthermore, the British Mandate White Book regulations of 1939 severely restricted purchase of land by Jews in the Galilee, Izre'el, and Negev regions. (17)

For these reasons, kibbutzim in the western and eastern parts of the Upper Galilee were allocated with relatively small amounts of water and small parcels of land, often removed from the settlements' immediate vicinity. (18) The situation was severe in this regard, and some kibbutzim were forced to seek sources of income other than agriculture. Many established small workshops and factories, and a few others later developed recreation facilities to supplement agriculture.

Unfavorable ecological conditions continue to exist outside the mountainous areas as well. In the southern part of the country (the Negev semiarid and arid areas), land has been available in sufficient quantities, but water resources have been extremely scarce, even after the construction of the National Water Carrier in 1964. This shortage is one of the factors which has compelled some kibbutzim in the south, most notably since the late 1960s, to seek industrial solutions at relatively high intensity industrial adoptions. (19)

Water and land shortages have not eased throughout the years. Many kibbutzim, especially in the south and in the mountainous regions, have only quite recently received only 70 percent of their planned water allocation. (20) The political need to establish new settlements made it necessary to divert some water allocation and land parcels from

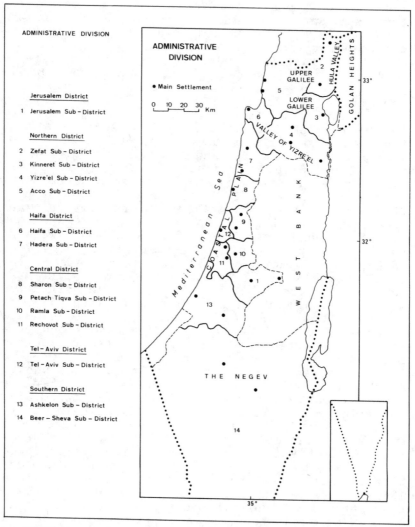

FIGURE 2 :   **ADMINISTRATIVE DIVISION AND GENERAL REGIONS OF THE
STATE OF ISRAEL.**

existing kibbutzim. (21) On the whole, the addition of new kibbutzim has always imposed further constraints on the agricultural expansion of existing kibbutzim.

Not only have land allocations declined through time, but population pressure on the agricultural land within kibbutzim has increased due to population growth, as revealed by Table 19.3. Until 1952 the land-population ratio increased due to the opening of new frontiers. However, another massive settlement wave occurred in the early 1950s, a considerable part of which took place within existing frontiers or within new frontiers with smaller relative amounts of cultivable land. Compounded by the factor of population growth, the overall land-population ration declined, and this trend continued through the 1960s and 1970s.

The amount of water available for agricultural purposes has also been declining. Table 19.4 shows the temporal trend of water allocation (in cubic meters) per irrigated land unit (in dunams). The two-decade trend shows a considerable decline of one-third in the amount of water available per irrigated dunam. Most of this decline is attributed to the fact that the increase in water allocation did not keep pace with the increase in the amount of land allocated for irrigated crops. Although intensive research in irrigation technology has been taking place, this could not provide a permanent solution to dependence on unsteady and declining resources. In fact, it has been argued that with present water-extraction technology, the amount of water available for agriculture will decline by a further one-third toward the end of the century.

Thus, the increase of population presure on agricultural land and the relative decline in the amount of water available for agriculture have put prospects for agricultural expansion in severe doubt, jeopardizing the economic survival of kibbutzim. Those kibbutzim which were located in poor environments were quite quickly compelled to search for other sources of income, primarily in industry and manufacturing, even though this involved a relaxation of their ideological principles.

Table 19.3. Cultivated Land-Population Ratio for all Kibbutzim, 1936-1972 (in dunams[*]).

Year	1936	1942	1947	1952	1957	1962	1967	1972
Land-Population ratio	9.63	10.35	6.63	19.58	16.40	15.91	13.87	11.68

[*] 1 dunam = 0.25 an acre

Source: Compiled from unpublished data of Audit Union of the Workers' Agricultural Society, Ltd.

Table 19.4. Water Allocation-Irrigated Land Ratio for all Kibbutzim 1953-1971.
(Cubic meters per dunam)

	1953	1957	1965	1968	1971
Water-land ratio	1,284	1,137	1,077	970	852

Source: Compiled from Israel Water Commission, Annual Reports.

Table 19.5. Agricultural Income for all Kibbutzim, 1936-1961 (Percentage).

	1936	1942	1949	1952	1957	1961
% agricultural income	62.6	60.7	58.8	67.6	76.8	75.9

Source: Compiled from unpublished data of Audit Union of the Workers' Agricultural Society, Inc.

Table 19.6. Employment in Agriculture for all Kibbutzim, 1936-1961 (Percentage).

	1936	1942	1949	1952	1957	1962
% agricultural employment	52.4	45.9	37.4	40.6	64.5	

Source: Compiled from unpublished data of Audit Union of the Workers' Agricultural Society, Ltd.

Economic Factors

Operating under water and land constraints, kibbutzim have been faced with two choices: to adjust to limited resources by attempting to intensify and stretch their utility, or to totally shift the economic emphasis away from agriculture. Kibbutz authorities realized that the whole essence of kibbutz life would be jeopardized by relinquishing agricultue, and those kibbutzim that could afford to, have retained their agricultural basis.

In the late 1940s this inclination was supported by a favorable governmental policy of agricultural expansion in order to supply food products for the rapidly growing population. At the same time, however, a strict import custom policy imposed by the government to encourage industrial development caused a rise in agricultural production costs. Furthermore, in the early 1950s the government attempted to maintain low food prices in order to hold down the rising cost of living. For welfare considerations, moshavim (new semi-communal agricultural settlements), which were the main competitors of kibbutzim, were favored in this policy. These factors, along with the imposition of output quotas, imposed hardships on kibbutzim. (23) Any expansion of agricultural activity would have been followed by a further relative decline in kibbutz profitability. (24) The inevitable result was a resort to other sources of income, reducing the relative share of agriculture.

These trends are supported by data in Table 19.5. Between 1936 and 1949 there was a decline in the relative share of agricultural income. The sharp increase toward 1957 is primarily due to the wave of new kibbutzim established in the early 1950s, almost all of which were totally dependent on agricultural income. However, the consequences of governmental intervention in the market in the early 1950s are demonstrated by the beginning of the decline in the role of agriculture in the late 1950s.

The relative decline in agricultural expansion that began to emerge in the mid-1950s due to resource constraints is also reflected in the trend of employment in agriculture (Table 19.6). Until 1957 relative employment in agriculture increased variably, but decreased significantly toward 1962 to its lowest level ever. Although part of the decline can be accounted for by technological progress,

some kibbutzim began to shift part of their manpower resources from agriculture to more promising
activities, primarily into industry. In fact, as
indicated above, the late 1950s marked the beginning of the accelerated stage of industry adoption
by kibbutzim.

Socio-Demographic Factors

In addition to constraints on physical resources
and and agricultural marketing opportunities, the
kibbutz economy underwent further pressure due to
the emergence of latent unemployment within the
kibbutz. This trend can be partially attributed to
the universal decline in the amount of labor needed
in agriculture due to technological process. However, the bulk of this trend can be attributed to
demographic developments within kibbutzim, namely,
population growth and changes in age distribution.
    Throughout the years, individual kibbutzim
have experienced considerable population increases.
This growth was due to natural increases and
immigration, the latter being composed of immigration from overseas and of Israeli urban youth
joining the kibbutz. Table 19.7, presenting the
growth in total kibbutz population between 1936 and
1973, reveals several trends. Firstly, total kib-
ᴜutz population grew nine-fold throughout the period. Much of the increase is due to the establishment of new kibbutzim, a trend that has slowed
considerably since the mid-1950s. This is reflected
by the more normal growth rates in the 1960s and
1970s expected from a relatively stable society.
Secondly a significant trend occurred between 1957
and 1962. As we have seen, the 1950s were times of
ɩelative decline in kibbutz profitability. Consequently, there was also a relative decline in the
standard of living in kibbutzim, resulting in
emigration and negative population growth. However,
the improvement in the standard of living in the
1960s curbed this trend and stabilized growth
rates.
    As a result of population growth, there have
been considerable changes in the size of kibbutzim
(Table 19.8). The percentage of smaller kibbutzim
(under 300 persons decreased almost constantly
throughout the period. The share of the medium-size
kibbutzim (300 to 700) increased, although incon-
sistently. However, the percentage of large kibbut-

275

Table 19.7 Total Kibbutz Population Growth, 1936-1973.

	1936	1942	1949	1952	1957	1962	1967	1973
Total Population	11,761	24,040	61,073	69,991	83,942	82,467	93,210	104,472
Mean Annual Growth Rate	–	20.9	25.7	4.9	4.0	-0.3	2.6	2.0

Source: Compiled from unpublished data of the Audit Union of the Workers' Agricultural Society, Ltd.

Table 19.8. Kibbutzim by population size, 1936-1973 (percentage).

	Under 200	200-299	300-399	400-499	500-599	600-699	700-799	800-899	900-999	1,000+
1936	53.4	20.9	6.9	2.3	6.9	2.3	6.9			
1942	42.8	32.9	12.0	5.4	1.1	2.2		3.2		1.0
1947	22.4	22.4	23.2	13.7	6.8	1.7	3.4	1.7		3.4
1952	37.3	16.3	12.1	14.9	8.4	4.2	0.4	1.8	0.8	3.2
1957	26.9	17.4	16.1	8.9	13.4	10.2	4.0	0.4	0.9	1.3
1962	23.7	16.8	16.8	12.5	12.5	7.3	7.7	1.2	0.8	1.2
1967	12.5	19.2	21.9	13.9	16.1	7.1	5.8	1.3	0.4	1.3
1973	8.3	12.0	25.1	17.6	13.0	12.0	4.1	4.1	1.3	1.8

Source: Compiled from unpublished data of Audit Union of the Workers' Agricultural Society, Ltd.

zim (over 700) increased only slightly as compared
to medium-size kibbutzim. On the whole, throughout
the period there has been a tendency of kibbutzim
to increase in size as witnessed by the fact that
toward the end of the period more than two-thirds
of the kibbutzim belong to the middle-size group.

Such a population increase, which in some
kibbutzim provided a considerable increase in human
resources, was not matched by an equivalent rate of
increases in physical resources for agricultural
expansion. As was earlier indicated, population
pressure on land increased with time, causing
latent unemployment. The pressure to provide productive jobs for able workers was great, and
industrial alternatives were sought in order to
reduce social stress.

This ecological-demographic problem was further complicated by the natural aging trend of the
kibbutz population. This factor began to affect the
labor resources in the 1940s and 1950s in older
kibbutzim, and in the 1960s and 1970s in relatively
younger ones. (25) The problem arose from the fact
that the physical ability of the aging people was
not equal to the demands of agricultural work, even
with high levels of mechanization. Since one of the
kibbutz's ideological principles calls for a guaranteed respectable occupation for each member for
as long as he can function, kibbutzim have been
faced with the social need to provide productive
but suitable jobs which suit the physical abilities
of their aging members. (26) Industrial occupations
were the most suitable and promising ones.

A further change in age distribution, which is
an outcome of population growth and size increase
in kibbutzim, is the growing size of second- and
third-generation populations. As part of the diffusion of modernization into kibbutzim, members of
the younger generation began to acquire higher
education in universities and technological insti-
tutes, especially during the 1960s and 1970s, and
have been willing to contribute to kibbutz life.
(27) In many cases, the acquired education could
not apply to the relatively narrow activities
within agriculture. In this case, kibbutzim were
also faced with the growing need to provide new and
rather modern industrial occupations that would
match personal and professional preferences.

Thus, for demographic reasons (population
growth and the growing number of aging members, on
the one hand, and young members, on the other

hand), kibbutzim inevitably searched for new areas of economic activity in which the demographic changes, as well as economic trends, could be best accommodated. The most promising areas, as shown earlier, were in industry and manufacturing.

CONCLUSIONS

If we adopt the concept of the production equation to comprehend the kibbutz as an agricultural production system, and if we discuss in its terms the previously described factors, we can understand the kibbutzim's departure from ideological principles in favor of adopting industry. All production factors remaining constant, the amount of actual production would positively depend on demand for the particular product. Increasing demand (growing national population) would have resulted in increased returns from agriculture, providing, in turn, capital for reinvestment and for raising the standard of living within the kibbutz. However, if agricultural output and marketing are constrained by institutional intervention and agricultural production factor availability grows more scarce and more expensive, per capita revenues decline. Coupled with population increase, this trend is followed by a decline in the standard of living, unless manpower is reallocated to other, more productive and less constrained economic activities. Thus, throughout the years kibbutzim have come to realize that agriculture alone cannot maintain the internal stability of kibbutz life. Industry was adopted as a tool to alleviate the social stress that resulted from the interaction of ecological, economic, and demographic problems. The differential exposure of kibbutzim to the influence of the various factors caused a differential and gradual removal of the resistance to adoption of innovative economic activity. The fact that the adoption of industry is shared by virtually all kibbutzim, and that kibbutz industry today constitutes an integral part of the national economic growth, substantiates the success of this economic venture.

REFERENCES

(1) E. Shoshani, The Kibbutz in Israel (Tel Aviv:
Hakibbutz Hameuchad, 1973 (In Hebrew)

(2) A. Shachal, "Kibbutz Industry: Development,
Structure and Problems" (Paper, University of Tel
Aviv, 1968). (In Hebrew)

(3) Ibid.

(4) H. Barkai, The Industrial Revolution on the
Kibbutz (Paper Presented at Van Leer Convention,
Tel Aviv, 1972). (In Hebrew)

(5) Kibbutz Industry Association, Ltd, Annual Reports: 1968-1976 (Tel Aviv, 1969-1977) (In Hebrew)

(6) Barkai, Industrial Revolution on the Kibbutz.

(7) H. Gan, "With or Without a Kibbutz Distinctive Industry," in Structural Changes in the Kibbutz Economy: A Symposium, Yad Tabenkin Publications, no. 3. (Tel Aviv: Hakibbutz Hameuchad, 1973)
(In Hebrew)

(8) C. Nahor, "Industrialization: A Practive
that Preceded Ideology," Hedim 30 (1965): 50-58.
(In Hebrew)

L. Sha'ashua, Y. Amiel, and Y. Goldsmith, The
Effects of Industry and External Services of Kibbutz Profitability in the 1960s Tel Aviv: Heshev,
1972) (In Hebrew)

(9) Kibbutz Industry Association, Ltd., Annual
Reports.

(10) Shoshani, The Kibbutz in Israel.

(11) Ibid.

(12) Ibid.

(13) A. Daniel, "The Israeli Case with Regard to
Interrelations Between the Government and the Cooperative Movement." In CIRCOM, Dynamics of Interrelations Between Agricultural Cooperatives and
the government: International Symposium. Tel Aviv,
1974.

(14) A. Meir, Diffusion Spread and Spatial Innovation Transmission Processes: The Adoption of Industry by Kibbutzim in Israel as a Case Study (Ph.D. Dissertation, University of Cincinnati, 1977).

(15) Y. Katzenelson, The Role of Precipitation in the Water Economy of Israel. Tel Aviv: Meteorological Service, 1969) (In Hebrew)

(16) Y. Karmon, The Land of Israel: Geography of the Country and its Regions (Tel Aviv: Yavneh, 1973). (In Hebrew)

(17) Encyclopaedia Judaica, "White Papers." Vol. 16 (Jerusalem: Macmillan, 1971).

(18) Z. Zur, "On Settlement in the Galilee" in Land and Water Policy: A Symposium, Yad Tabenkin Publications, No. 16 (Tel Aviv: Hakibbutz Hameuchad, 1976). (In Hebrew)

(19) Y. Va'adia, "Settlement in Arid Areas," in Land and Water Policy: A Symposium, Yad Tabenkin Publications, No. 16 (Tel Aviv: Hakibbutz Hameuchad, 1976). (In Hebrew)

(20) Agricultural Workers' Union, The Agricultural Union and Its Activities: 1970-1973 (Tel Aviv: Report to the 12th Convention, 1974) (In Hebrew)

(21) D. Raviv, "Legal Aspects of Agricultural Land" in Land and Water Policy: A Symposium, Yad Tabenkin Publications, No. 16 (Tel Aviv: Hakibbutz Hameuchad, 1976) (In Hebrew)

(22) E. Kally, "Long Range Water Economy," In Land and Water Policy: A Symposium, Yad Tabenkin Publications, No. 16 (Tel Aviv: Hakibbutz Hameuchad, 1976). (In Hebrew)

(23) A. Argov, "The Kibbutz in the Economic Struggle" MiBifnim 19 (1956): 26-36. (In Hebrew)

(24) Y. Don, Industrialization in Advanced Rural Communities (Paper presented at the Fourth World Congress of Rural Sociology, (Turon, Poland, 1976).

(25) U. Leviatan, Work Life for the Older Person on the Kibbutz (Givat Haviva: Center for Social Research on the Kibbutz, 1975).

E. HaCohen, "Trends in the Demographic Development of the Kibbutz" Hedim 24 (1959): 18-25 (In Hebrew)

(26) D. Atar, "Industrialization and Elderly Labor in the Kibbutz" in S. Shur, ed., The Kibbutz as a Distinctive Society (Tel Aviv: Hakibbutz Haartzi, 1975) (In Hebrew)

S. Shachar, "Employment Problems of the Elderly on the Kibbutz," Hedim 33 (1968): 108-121. (In Hebrew)

(27) N. Shiran, "On Integration of Agriculture and Industry," in Structural Changes in the Kibbutz Economy: A Symposium, Yad Tabenkin Publications, No. 3 (Tel Aviv: Hakibbutz Hameuchad, 1973) (In Hebrew)

V
Agro-Industrial Integration and Rural Transformation

20 Non-metropolitan Industrial Employment and Rural Transformation in the United States

Richard Lonsdale

Rural transformation in the United States has come to mean a fundamental change in the employment base, most notably a shift from farm employment to that in decentralized manufacturing and, more recently, to that in the services sector. In rural and small-town (non-metropolitan) (1) areas, farm employment is now overshadowed by factory (2) employment by a margin of almost two to one. Not surprisingly, an important part of this manufacturing is agro-industrial in character: processing of food and feed products, agricultural equipment and supplies, fertilizers and other chemicals, and the assembly of a wide variety of goods for consumption in nonmetropolitan areas. Modern agriculture, with its seemingly ever-greater productivity per farm worker, has, of course, served as a force pushing labor off the farms. Decentralized manufacturing, meanwhile, has expanded employment at a pace more than offsetting farm employment declines, thus providing a net expansion of basic employment in rural and small-town areas.

Rural transformation in the United States has also meant a fundamental demographic transformation. The long-established pattern of net population out-migration from nonmetropolitan areas was slowed in the 1960s, and then reversed. (3) Rural and small-town areas, including those rather remote from larger cities, are now increasing their population at a rate exceeding that for metropolitan areas. (4) With decreased out-migration and increased in-migration, rural and small-town areas

are now experiencing net in-migration. This "population turnaround" was preceded by the large-scale
expansion of decentralized industry and the expansion of non-farm employment opportunities. New
industrial employment and demographic change have
thus been closely interrelated in promoting an
overall rural transformation.

This book focuses on "modern agriculture," a
theme here treated as embracing not only those
activities traditionally seen as purely agricultural in character, but also including such func-
tionally related and critically supportive activities as agro-industrial and other manufacturing
operations located within the rural sphere. As
previously noted, this manufacturing has been directly associated with fundamental changes in employment and population trends. And if this manufacturing is seen as an integral part of modern
agriculture, then it is perfectly plausible to
contend that in the United States modern agriculture has had a profound effect upon rural transformation, bringing about a degree of modernization
that has seen rural areas achieve a quality of life
on par with or ahead of that in urban centers in
many essential respects.

The problems of rural non-metropolitan areas
in all advanced societies have traditionally been
those associated with a real concentration and
regional inequity. As a nation industrializes,
there follows a technological modernization of
agriculture, bringing increased productivity and a
declining need for farm labor. In the United
States, this occurred shortly after the farmlands
were first settled and cultivated. As larger
urban-industrial centers emerged, with expanding
employment opportunities, the migration of people
from rural to urban places was a natural consequence. Net out-migration of people, especially
those younger and with more education, became a
standard feature of rural and small-town areas. As
a result, such areas commonly suffered the combined
problems of declining population, limited employment prospects, loss of critical central-place
functions (such as retail stores, post offices,
schools, and medical facilities), and a general
lack of confidence in the economic and social
future of the area. (5) The "better life" and the
jobs were concentrated in the urban centers, and
regional inequity seemed to be an inevitable

by-product of industrialization.

Today much of this has changed in rural and small-town America. As a whole, these areas are now enjoying a net in-migration and a population growth rate exceeding that of the nation as a whole. Residential preference polls show that small towns are preferred (over large cities) as places to live. (6) The pace of industrial job expansion greatly exceeds the national average. A renewed sense of confidence is clearly evident in small-town business areas and in the construction of new homes. Those amenities once thought to be available only in the city are now thought to be available there to the same or greater extent than in the city. It has been a remarkable transformation, all in about twenty-five years. What have been the key ingredients? The following comments attempt to provide a theoretical as well as an applied framework for viewing the transformation.

THE CONCENTRATION-DECONCENTRATION THESIS

It can be reasoned (and the record in some advanced industrial societies provides support) that nations go through a cycle of geographic inequity. Prior to industrial revolution, much of the population is agrarian, most of the industry is of a cottage type, and there are comparatively few reasons (other than ones relating to the inherent natural quality of the land for agriculture) for regional inequalities. However, the advent of modernization and concomitant industrialization and urbanization quite obviously promote geographic inequality and inequity (the latter term incorporates the notions of "fairness" as well as "inequality"). Industry tends to concentrate in the new and growing urban centers to take advantage of scale and agglomeration economies and of improved access to markets and suppliers. (7) Supporting the tendency to concentrate geographically, some analysts reason, are those improvements in transportation which encourage manufacturers to locate in a few places and focus their attention on reducing production costs through large-scale operations. (8) In this process, less urbanized rural and small-town areas are "left behind."

Eventually, it is theorized (and the United States experience supports the theory), cities become too big, diseconomies of scale set in, city

images deteriorate, and deconcentration is encouraged by economic factors, government policy, and
public attitudes. Regional inequities commence to
diminish. (9) Transportation continues to play a
critical role by providing industry with comparatively inexpensive access to the national market
from almost any location, thanks (in the case of
the United States) to widespread auto ownership and
trucking services and to an excellent network of
highways. (10) Costs of short transport hauls decline relative to those for longer hauls. Industry
decentralizes and, of course, employment opportunities and population patterns follow suit.

It would seem, therefore, that industrial
societies go through a phase of geographic concentration (of industry, jobs, and people) followed by
a phase of deconcentration. The United States is
now very much in the deconcentration phase. Plotted
on a graph, the cycle traces out an inverted "U"
curve. (11) But a haunting thought in all of this
is the idea that "inexpensive" energy as reflected
in transportation costs has made it all possible.
It remains to be seen what the geographic impact
(on jobs and people) will be of expected higher
energy costs and energy shortages.

THE RECENT TREND TO INDUSTRIAL DECENTRALIZATION IN THE UNITED STATES

Despite a lack of comprehensive planning and coor-
dinated policy to achieve it, the United States has
witnessed a remarkable degree of industrial decentralization. In the early 1950s an "industrial
turnaround" was achieved; that is, the century-long
trend toward the geographic concentration of manufacturing was reversed and the decentralization
phase began, consistent with the concentration-
deconcentration thesis discussed previously. In the
United States, this "turnaround" was the combined
result of many factors, among them the expansion of
trucking able to serve virtually every community;
the interstate highway system; almost universal
automobile ownership, permitting rural residents to
commute daily to even moderately distant plants;
rural-urban wage differentials; the recognition
that rural labor, while not as skilled, had a
strong work ethic; the increased "routinization" of
many production procedures; the desire by management to escape the high taxes and strong labor

unions characteristic of larger cities; the growing
disenchantment with the quality of life in cities;
growing regional markets; and the feeling that
industry was "more appreciated" in less urbanized
areas. (12) Such a powerful combination of decentralization motives has not been present in other
advanced societies, and thus United States manufacturing has come to be more extensively decentralized than manufacturing in other nations, despite the United States being one of the few
advanced nations to have no real policy toward this
end.

In this study, the share of the nation's total
manufacturing employment found in non-metropolitan
(rural and small-town areas) is used as a measure
of the level of a real concentration/deconcentra-
tion. At the time of the "industrial turnaround,"
about 20 percent of all United States manufacturing
employees were in such non-metropolitan areas.
(13) In the late 1950s, deconcentration tendencies
began to gather strong momentum, and between 1962
and 1978, the share of United States industy in
non-metropolitan areas climbed from 23.5 to 28.8
percent (Table 20.1). In the 1962-1978 period
non-metropolitan manufacturing employment increased
by 1.8 million, or 47 percent, compared with a
metropolitan increase of 1.4 million, or 11 percent. (14) Nonmetropolitan areas thus accounted for
56 percent of the national net gain in industrial
jobs, an impressive feat for that section of the
nation with but 31 percent of the population in
1970.

The industrialization of the American countryside is no longer a vision for the future. With 31
percent of the population and 29 percent of the
factory employment in non-metropolitan areas, it is
now a reality. The United States is apparently the
only advanced society with a reasonably large
territory to be able to claim this. Indeed, if
current trends persist, non-metropolitian areas
will soon become more industrialized than metropolitan areas, (15) leading one to speculate about the
possibility of a space economy consisting of islands (large cities) providing service functions
surrounded by a sea (rural and small-town areas)
focusing on agriculture, mining, and manufacturing.
But once again, the question of energy cost and
availability casts a shadow over such an extrapola-
tion.

	Total U.S. (thou.)	Metropolitan[b] (thou.)	(thou.)	Nonmetropolitan (percentage of U.S. total)
1959[c]	16,173	12,792	3,381	(20.9)
1962	16,622	12,715	3,907	(23.5)
1967	19,390	14,541	4,849	(25.0)
1969[c]	20,177	15,435	4,742	(23.5)
1970	19,764	14,654	5,110	(25.9)
1974	19,983	14,320	5,663	(28.3)
1978	19,870	14,141	5,729	(28.8)
Change, 1962-78	3,248	1,426	1,822	

Source: Derived from detailed data given by Claude C. Haren and Ronald W. Holling, "Industrial Development in Nonmetropolitan America: A Locational Perspective," In Nonmetropolitan Industrialization, ed. R. E. Lonsdale and H. L. Seyler, (New York: John Wiley, 1979), p. 18; and Claude C. Haren, "Location of Industrial Production and Distribution," in Rural Industrialization: Problems and Potentials, ed. L. R. Whiting (Ames: Iowa State Univ. Press, 1974), p. 8.

[a]Wage and Salary employment, adapted from Bureau of Labor Statistics--Employment Security estimates. All data for March of year indicated.

[b]Includes 225 mostly larger of 278 metropolitan areas designated through 1977.

[c]Data adapted from Country Business Patterns and thus not fully comparable with data for other years.

286

EXPANDING THE RURAL EMPLOYMENT BASE

The role of industrial employment is here considered within the context of standard economic base (or export base) theory. It is reasoned that the export sector of the local economy provides the basic employment, which in turn supports the local population through the importing of capital. The basic or "community-forming" activities are thus ones where the final product is exported out of the area. The non-basic or "community-serving" activities provide goods and services to the local area.

Most non-metropolitan manufacturing, even that of an agro-industrial nature, is basic in that most of the products leave the immediate area (for example, the county where the plant is located). Thus, for each new manufacturing job, there is a presumed increase in non-basic employment, and thus a multiplier effect. In most cases, the employment multiplier has been between 1.0 and 2.0 (1.0 signifies no net increase beyond the number of new industrial jobs, while 2.0 indicates that fifteen new industrial jobs will cause a total increase of thirty in local employment). In a summary of twelve existing non-metropolitan case studies, Summers and his associates found employment multipliers between 1.0 and 1.7. (16) Stevens and Wallace, in a 1947-1960 study in Indiana, found an employment multiplier of 1.4. (17) In a Utah county from 1950 to 1966, McArthur and Coppedge found a multiplier of 1.7. (18) In Missouri, Braschler found manufacturing-caused long-run employment multipliers averaging about 1.6 around 1950, 1.9 around 1960, and 2.1 around 1970; thus, he proposes that the multipliers become larger through time. (19) Stronger multipliers are associated with new industries that pay good wages, buy at least some of their materials and services locally, and employ local workers who do most of their shopping locally.

In non-metropolitan America as a whole, industrial expansion clearly played a critical and decisive role in expanding the employment base in the 1960-1970 period. The absolute gain of 1.25 million industrial jobs represented 31 percent of the 4.06 million total gain in non-agricultural employment. (20) Assuming that most manufacturing is basic in character, and assuming at least a modest (say, 1.5) employment multiplier, manufacturing probably accounted for mearly two million new jobs. This far exceeded the estimated 1.12

million decline in farm employment in the same
period, (21) and the non-metropolitan employment
base thus diversified and expanded to a very
impressive extent.

There are, of course, other kinds of basic,
job-generating activities that rural and small-town
areas might acquire, such as those associated with
tourism, recreation, retirement, mining, expanded
agriculture, government facilities, and transportation. Employment in these sectors has been growing
much more rapidly than in manufacturing. But the
great majority of rural and small-town areas lack
the warm winters, mountain scenery, special situa-
tional advantages, or natural resources to attract
any significant amount of new basic activity other
than manufacturing. Manufacturing may not be a
growth sector of the national economy, but it has
demonstrated a remarkable degree of locational
mobility.

In the 1970-1978 period, manufacturing accounted for 619,000 new jobs in non-metropolitan
United States, only 14 percent of the total non-
metropolitan non-agricultural employment gain of
4.6 million. (22) Clearly, the focus has now
shifted to employment gains in the tertiary or
services sector (such as trade, personal and professional services, finance, transportation, communications, and government). It can be reasoned
that this shift in employment emphasis is perfectly
logical. In effect, strong gains in service employment have followed strong gains in manufacturing after a certain period of time (five to ten
years). A useful framework for viewing this is
provided by what might be called a primary-second-
ary-tertiary transitional thesis. As a nation or
region achieves economic growth, agriculture declines in relative importance, giving way to manufacturing. Then, in time, manufacturing expansion
levels off, accompanied by growth in such service
sectors as wholesale and retail trade, personal and
professional services, finance-insurance-real estate, and government. This long-term shift in
emphasis from primary (agriculture) to secondary
(manufacturing) to tertiary (services) activity is
characteristic of maturing economies.

DEMOGRAPHIC AND LIFE-STYLE IMPACTS

The impact of the expanded and diversified employment base in rural and small-town America is widely evident, particularly as it has involved population levels and life-styles. Indeed, one needs only to drive through the small towns of the South, the Middle West, the Great Plains, and other areas of the United States to appreciate the impact. Thousands of smaller communities - after years of population loss, economic stagnation, abandoned retail shops, and general lack of confidence in the future - are showing physically tangible signs of renewed prosperity and growth. This is evident in the new housing; spruced-up retail shops; fast-food outlets; active automobile and farm implement agencies; modern banks, schools, and post offices; improved sewerage, water, and recreational facilities; and, almost invariably, one or more new industrial plants on the edge of town. Those with a nostalgic bent often argue that small towns are losing some of their traditional charm. This is probably true, but the return of at least a modest rate of growth - after many decades of dormancy or decline - has been enthusiastically received by most people throughout rural America. (23)

The all-essential element has been a resumption of population growth in small towns and rural areas. Since the 1970 census (and probably a few years before that), nonmetropolitan areas have enjoyed a net in-migration of population and an overall growth rate exceeding that of metropolitan areas. At first focused on "adjacent" nonmetropolitan areas (those near metropolitan areas), it soon came to include the more distant ("nonadjacent") nonmetropolitan districts. (24) Increased in-migration, as well as reduced out-migration, helps to explain the population growth. In a study of migrant response in four rural areas in the 1965-1970 period, Olsen and Kuehn found that 22 percent of the new industrial jobs were held by migrants, including former residents who elected to return. (25) In the Great Plains, Kale found that 21 percent of the employees at male-majority plants moved into the area to take their new jobs, while the figure was 6 percent at female-majority plants. (26)

Many factors influence persons' decisions to migrate or remain where they are, but employment

residents, for the most part, enjoy the same set of housing comforts that city residents do. Income levels are lower, but an important part of this difference is balanced by a lower cost of living. Employment opportunities are greater, and many people, in particular those just out of school who might well have out-migrated to the city twenty years ago, are now electing by their own free will to remain in the areas where they grew up, near their families and long-time friends. The social and psychological significance of this - in a nation with so much geographic impermanence in individual residential behavior - should not be underestimated.

CONCLUSIONS

The expansion of industrial employment in rural and small-town areas of the United States should be seen as a normal and essential phase in the overall modernization and economic and social transformation of these areas. Forces common to many advanced societies have been operative, (33) but a combination of circumstances has seen the United States as a pacesetter in rural transformation. Other nations may experience an essentially similar change. Rural America passed through a natural phase of a real concentration of industry and people, and now an equally natural trend to deconcentration is widely evident. In a sense, rural and small-town areas have been "developing lands" transcending agrarian emphases and moving on to manufacturing and subsequently to service-performing activities. The 1960s saw the peak of the industrialization phase, and now the emphasis has shifted to the services sector. With this transformation, rural and small-town residents have come to enjoy employment opportunities, the availability of services, housing, and a life-style similar in many qualitative respects with those enjoyed by city residents. Nonmetropolitan industrialization, inexorably linked with and a part of the modernization of agriculture, has wrought a fundamental transformation of the American rural scene. Other societies, less advanced in this rural transformation, can learn much from the United States experience.

opportunities are certainly a dominant consideration. (27) Most studies support the almost
self-evident proposition that new jobs have a
positive population impact. In a survey of 58
existing case studies of small-town industrialization, Summers and his associates found that 86
percent subsequently experienced a population gain.
(28) In Dietz's northern Great Plains study, area
population decline was reduced and towns grew more
impressively in thirteen counties receiving industry, compared with twenty-five counties which did
not. (29) In a Missouri study of rural and
semi-rural counties, Braschler found a population
multiplier of about 5.0; that is, 100 new industrial jobs brought a population increase of 500. (30)

New industrial jobs do not necessarily mean an
improvement in per capita income levels (31) (although income levels generally do improve in
lower-income areas, especially in the South), but
they do signal an expanded employment base, more
families with two wage earners, and thus higher
absolute income levels in the community. This
translates into more retail sales and improved
retail facilities. It also translates into a more
active real estate market, higher property values,
increased local tax revenues, and improvements or
expansions in public services. In effect, small
towns assume more of the characteristics of larger
towns.

Life-styles in small towns and rural areas,
while likely to remain distinct in a number of
essential attributes, are becoming more and more
like those in the cities. (32) Increasingly, these
distinctions are being viewed as assets by rural
Americans rather than liabilities. Many of the
traditional deficiencies of rural life, social and
economic, have been reduced or eliminated. Thanks
to the national news media, rural and small-town
residents have come to embrace a set of attitudes
and aspirations similar to the nation as a whole.
Good shopping facilities may be ten or even forty
miles distant; but with one or two automobiles in
the family, that is no serious problem. Educational
facilities, while still below the national norm,
are not dramatically different in quality, and
going off to the state university after high school
is as commonplace as elsewhere. The days of outdoor
privys and loading the furnace on a cold winter
morning are essentially gone; nonmetropolitan

NOTES

(1) "Nonmetropolitan," roughly comparable to rural and small-town areas, refers to United States areas outside of counties, or groups of contiguous closely-tied counties, which contain at least one city of 50,000 population or more (i.e. metropolitan areas or SMSAs).

(2) The terms "factory," "industrial," and "manufacturing" are here used synonymously, as has become common practice in the U.S.

(3) Calvin L. Beale, The Revival of Population Growth in Nonmetropolitan America, ERS-605 (Washington: U.S. Dept. of Agriculture, Economic Research Service, 1975).

(4) Glenn V. Fuguitt and Calvin L. Beale, Population Change in Nonmetropolitan Cities and Towns (Washington: U.S. Dept. of Agriculture, Economic Research Service, 1976).

(5) Richard E. Lonsdale, "Background and Issues," in Nonmetropolitan Industrialization, R.E. Lonsdale and H.L. Seyler, eds. (New York: John Wiley, 1979), pp. 5-9.

(6) Glenn V. Fuguitt and James J. Zuiches, "Residential Preferences and Population Distribution," Demography, 12, 3 (1975): 491-504.

(7) James A. Chalmers and Michael J. Greenwood, "Some Thoughts on the Rural to Urban Migration Turnaround," International regional Science Review 2,2 (1977): 167.

(8) Benjamin Chinitz, "The Effect of Transportation Forms on Regional Economic Growth," Traffic Quarterly 14 (1960): 129-142.

(9) Jeffrey G. Williamson, "Regional Inequality and the Process of National Development," Economic Development and Cultural Change 13, 4, 2 (1965): 8-10.

(10) Chinitz, "Effect of Transportation Forms"; Gerald Kraft et al., The Role of Transportation in Regional Economic Development (Lexington: D.C. Heath, 1971).

(11) Williamson, in "Regional Inequality," hypothesizes this curve, but does so with reference to regional income inequalities rather than manufacturing.

(12) Steven R. Kale and Richard E. Lonsdale, "Factors Encouraging and Discouraging Plant Location in Nonmetropolitan Areas," in Nonmetropolitan Industrialization, ed. R.E. Lonsdale and H.L. Seyler (New York: John Wiley, 1979), pp. 47-55.

(13) Daniel Creamer, Changing Location of Manufacturing Employment, Part I (New York: National Industrial Conference Board, 1963), pp. 30-31, 130-131.

(14) Claude C. Haren and Ronald W. Holling, "Industrial Development in Nonmetropolitan America: A Locational Perspective," in Nonmetropolitan Industrialization, ed. R.E. Lonsdale and H.L. Seyler (New York: John Wiley, 1979), p. 16.

(15) Preliminary data from the Economics, Statistics and Cooperatives Service, U.S. Dept. of Agriculture, suggest this may already be the case at the present time (1980).

(16) Gene F. Summers et al., Industrial Invasion of Nonmetropolitan America (New York: Praeger, 1976), pp. 55-56.

(17) J.B. Stevens and L.T. Wallace, Impact of Industrial Development on Howard County, Indiana, Bulletin 784 (Lafayette: Purdue University Agricultural Experiment Station, 1964).

(18) J.W. McArthur and Robert Coppedge, "Employment Impacts of Industrial Development: A Case Study of Box Elder County, Utah," Utah Economic and Business Review 29, 2 (1969): 1-10.

(19) Curtis Braschler, Regional Growth in Missouri, SR 153 (Columbia: University of Missouri Agricultural Experiment Station, 1973), pp. 11, 33.

(20) Claude C. Haren, "Location of Industrial Production and Distribution," in Rural Industrialization: Problems and Potentials, ed. L.R. Whiting (Ames: Iowa State University Press, 1974), p. 8.

(21) Ibid.

(22) Haren and Holling, "Industrial Development,"
p. 18.

(23) One example of this attitude is provided by
R.E. Lonsdale, J.C. Kinworthy, and T.R. Doering,
Attitudes of Manufacturers in Small Cities and
Towns of Nebraska (Lincoln: Nebraska Dept. of Eco-
nomic Development, 1976).

(24) Fuguitt and Beale, Population Change.

(25) Duane A. Olsen and John A. Kuehn, Migrant Response to Industrialization in Four Rural Areas,
1965-1970, Agricultural Economic Report 270 (Washington: U.S. Dept. of Agriculture, Economic Research Service, 1974), p. 3.

(26) Steven R. Kale, Labor Supplies for Nonmetropolitan Manufacturing Plants in Small Communities
of the Central Plains (Lincoln: Nebraska Dept. of
Economic Development, 1979), p. 16.

(27) Curtis C. Roseman, Changing Migration Patterns within the United States, Resource Paper 77-2
(Washington: Association of American Geographers,
1977), pp. 4-9.

(28) Summers et al., Industrial Invasion, p. 21.

(29) John Dietz, "Rural Area Development: Analysis
of the Impact of New Factories on Agricultural
towns in the Northern Great Plains," Great Plains-
Rocky Mountain Geographical Journal 1 (1972): 24.

(30) Braschler, Regional Growth, p. 15.

(31) H.L. Seyler, "Industrialization and Household
Income Levels in Nonmetropolitan Areas." in Nonmetropolitan Industrialization, ed. R.E. Lonsdale and
H.L. Seyler (New York: John Wiley, 1979), pp.
149-159.

(32) For a general treatment of this theme, see
several papers in Thomas R. Ford, ed., Rural
U.S.A., Persistence and Change (Ames: Iowa State
University Press, 1978).

21 The Impact of Agro-industrial Development on the Viticulture of Germany: The Case of the Kaiserstuhl Region

Jorg Stadelbauer

THE PROBLEM

In the Federal Republic of Germany, viticulture belongs to those specialized agricultural systems in which agro-industrial integration has been carried through earlier than in other fields of agricultural production. The latest changes have affected, above all, the stages of production following the agricultural production (grape pressing, wine making, aging) and the meliorative basis of the cultivation (slope-terracing, vineyard consolidation). The position of the vine growers has been influenced in such a way that the alteration concerning the preceding and following stages are, at the same time, an expression of social change.

Regarding production, the interconnected elements of the wine industry can be seen as a viticultural complex. This chapter is an attempt to show some of the characteristics determinating vine-growing regions as spatial viticultural complexes from the point of view of agricultural and economic innovation. For the focus of our examination, the Kaiserstuhl area has been chosen as an example of such innovation. (1) It is a hilly region in the southwest of Germany, distinctly marked off from the surrounding area.

The theoretical framework of this chapter is formed by the conception of agro-industrial complexes in such a way as the concept has been developed by Soviet agricultural science in regard to an economic and sociocultural system completely

differing from the West German system. (2) The
critical discussion about different facts arising
from this concept is not to be ignored, (3) for it
too is a constitutive part of the theory and of the
practical application of theoretical knowledge.

Nevertheless, this chapter must be restricted
to a very few problems which appear to be apt to
stress the categorical basis of agro-industrial
integration: the normative fixation by juridical
acts; requirements of space for utilization; overcoming of space by economic connections; temporal
and social changes; economic raising of productivity by concentration and specialization, by a
cooperation between the different branches and
enterprises, and by agricultural, technical, and
economic innovations.

THE VINEYARD CONSOLIDATION
IN THE INTEGRATION PROCESS

Today, modernization in a winegrowing area finds
its most striking physiognomical expression in the
consolidation of vine holdings. In the Kaiserstuhl, it is connected with the arrangement of huge
terraces on which a modern growing of quality wine
takes places. The terracing of slopes in the
Kaiserstuhl corresponded with an extremely scattered pattern of holdings as long as gavelkind
tenure was applied. The modest technical equipment
of the past centuries made earth movements on a big
scale impossible, so the terraces mostly followed
the natural relief. Proportionate to the degree to
which modern agricultural techniques were introduced in the wine growing area, it became necessary
to enlarge the holdings in order to make them more
profitable. That is why land consolidation was
started after World War II. It corresponds to a
comprehensive organization of land which comprises
a bundle of measures for the improvement of agricultural structures: transformation of the terrain
by extensive earth movements; hydraulic measures;
improvement of the road system within the region;
fixation of a plan for vine plantation; and a new
regulation of the spatial attachment of ownership
and property rights. The juridical basis is the
Land Consolidation Law of 1976 (Flurbereinigungsge-
setz), which intends to improve the production and
working conditions in agriculture and which empha-

sizes, more than the older statute from July 14, 1953, the basis for previous proceedings – the protection of the environment.

The procedure of these measures shall be presented with the example of land-consolidation proceedings on a small scale in the northwest of the Kaiserstuhl (about forty hectares). In the mid-1960s, the community of Leiselheim appealed for land consolidation. The implementation was delayed because hydraulic measures (building of a retention basin) had to be prepared, approved, and started. In addition, the huge, planned earth movements made it necessary to improve the layer of loess by provisions of material from a neighboring forest. In July 1972, a hearing of the participating organizations and authorities took place as set out in the Land Consolidation Law. The arising differences of opinion had to be clarified in the following two and a half years. The official directions issued on January 31, 1975, made the start of the procedure possible. Further management was undertaken by the Administration for Land Consolidation and Settlement (Landesamt fur Flurbereinigung und Siedlungswesen). At the same time, the landowners' association was founded, and it existed as a corporation of public law (according to juridical regulations) for the duration of the consolidating procedures.

Simultaneously, the establishment of plans for roads and waters preceded the terrain measures, which were carried through within three months (February to April, 1976). While the new vines were planted according to the plan worked out by the regional government (Regierungsprasidium) in cooperation with the Institute for Wine Industry (Weinbauinstitut), the winegrowers' cooperatives, and the commercialization establishments, the deep furrow work and road building continued until the end of the year.

In the spring of 1977, one year after the new plantation was initiated, the statutes of ownership were newly regulated in a provisional form. They payed regard to the old ownership structures and the results of land evaluation. A definitive allocation of proprietor's titles according to the listing in the land register has been completed by 1979.

These proceedings clarify that modern land consolidation is accomplished by different author-

ities that have to act as mediators between opposing interests and opinions concerning land utilization and protection. The juridical basis ensures
that those who are personally affected by these
measures have the right to participate in the
different steps of the procedure. Participators'
associations and vine-planting cooperatives characterize a strong cooperative movement. The use of
cross-country machines and photogrammetric topographic surveys emphasize the claim to a very
precise way of working. The accompanying building
measures (roads, waters) necessitate cooperation
with other administrative and economic branches.
These facts show that the total complex of land
consolidation is an important part of integration
in regard to the wine industury and to area
planning.

In the years 1950 to 1974, sixty-three proceedings of public land consolidation in the
Kaiserstuhl (representing a total of 3,400 hectares) were carried through. A further 250 hectares
were consolidated voluntarily without public works.
The allotment on the huge terraces led to exactly
regulated bundles of strip-shaped parcels. These
regulations reduced the amount of work per hectare
by nearly half - to intensify winegrowing and to
increase its profitability. But the consequences of
these large-scale alterations in the landscape have
not always been positive. Negative impacts are
mainly in the climatic sphere. Recently, micro-
climatological topographic surveys and interpreta-
tions of test photographs taken with the remote
sensing scanner have proved that the small anti-
thetic inclination of the terrace surface favors
the formation of small, cold air cells and secondary wind systems. This means an increased danger of
frost. From the consumer's point of view, it is
often criticized that the traditional individuality
of small sites disappears. Improvements in the
winemaking and in the aging resulted in a standardization of quality. The measures of land consolidation additionally assure those fixations which have
to be made on the size of the cultivated land on
the basis of the Wine Industry Law (Weinwirt-
schaftsgesetz) of May 9, 1968; only suitable areas
which guarantee the growing of quality wine are
included. For the Kaiserstuhl, this means that only
those terraced slopes which are situated below 395
meters above sea level can be cultivated.

Altogether, the land consolidation shows a govenmental activity with a strong spatial impact; that is, the effect of legal standardization on the modern change of agricultural landscape. The individual measures are controlled by a comprehensive system that is meant to promote, with political instruments, the growing of quality wine to guarantee the profitability of enterprises and at the same time to accelerate the economic integration.

TYPES OF WINE ENTERPRISES
IN THE INTEGRATION PROCESS

The winegrowing complex "Kaiserstuhl" is characterized by the juxtaposition of several types of enterprises, which, to various extents, take part in cooperation and integration. With regard to the whole winegrowing area of Baden, they differ remarkably from the other regions according to their share in the wine industry: the cooperative sector holds 80 percent, the winegrowers' and a wine dealers' sector as well as the estates hold 5 percent, and the wine merchants' firms hold approximately 10 percent. In addition, the types of enterprises differ from each other as to their locational characteristics, their spatial connections, and their production profiles (see Table 21.1).

Winegrowers

The winegrowers' operations are based on family labor without additional workers. As a rule, the winegrowers are members of a cooperative that undertakes the making and the commercialization of the wine, or they are contractual suppliers of several large winegrowing estates or wine merchants' firms. The degree of specialization varies to a certain extent. If agriculture is the main source of living in principal income units, mixed types of enterprises are common. For subsidiary farmers, such as in the central Kaiserstuhl, whose main nonagricultural occupation is mostly somewhere outside the place of residence, winegrowing is often the only agricultural source of income. The farm size structures also differ. In the last two decades, the smallest properties of less than one hectare (mostly farmers with additional nonagri-

Table 21.1 Characteristics of the main types of enterprises in the Badish viticulture.

Type of enterprise	Location	Number	Relation	Integration	Function
Central Winery	Central location in regard to the Badish viticulture, peripheral location as to the Kaiserstuhl	1	Catchment area: Baden Region of sale: Mainly Baden-Wurttemberg, depots in other German regions, share of exportation 4-5% (enlargement planned)	Strong concentration on cooperative basis	Propaganda Commerce Wineaging Winemaking Reception of grapes Purchase of auxiliary means
Winegrowers' cooperative	Located in villages with a high percentage of viticultural land utilization; dispersion according to the network of group settlements	20	Catchment area: one community's territory (75% of the grapes at least have to come from this area); Region of sale: mainly Baden-Wurttemberg (wholesale-nutrition sector of the retail sale, sale to private persons); little exportation to other German regions, nearly no exportation to other countries	Concentration and specialization on cooperative basis	Propaganda Commerce Wineaging Winemaking Reception of grapes Purchase of auxiliary means
Wine estate Winegrower selling his own production	Mostly located in the greatest villages (Endingen, Ihringen), influenced by the historical development of the farm size structure	approx. 70	Analogous to the winegrowers' cooperative, but sales are mostly confined to Baden-Wuerttemberg	Individual enterprises; relations to other economic partners based on contracts or free	Propaganda Commerce Wineaging Winemaking Reception of grapes Purchase of auxiliary means
Vintager	Little farmers in all villages, knowing the gavel-kind tenure in the past (Kleinbauerntum)	approx. 2,350	As farmers connected to their parcels, as vintagers to the winegrowers' cooperatives in their villages, no wine	Party specialized. party diversification, cooperation with the winegrowers' cooperative or (seldom) wine estate	Propaganda Commerce Wineaging Winemaking Reception of grapes Purchase of auxiliary means

cultural income) and medium-size properties with more than ten hectares (if combined with winegrowing where agriculture is the principal income) experienced a relative growth, whereas the share of farms with sizes in-between has fallen. This distinction corresponds to a development that is noticeable in the whole agriculture of West Germany.

Winegrowers and Dealers

Individual wine making by wine growers and dealers is nowadays only efficient if the size of the vine holding is at least 1.5 hectares. As small properties dominate in the Kaiserstuhl, the number of such enterprises is small. In addition, the access to the trade network is limited. Wine estates play an equally minor part in the Kaiserstuhl, differing from the vine-growing areas on the Rhine and Mosel. However, they do exist occasionally. They unite the locational qualities and the organiztion of individual holdings and winegrowers' cooperatives. In some cases, the number of supplying vintagers runs into several hundreds and the wine estates are then similar in size to the cooperatives. While the wine estates have their own winegrowing, the wine trading firms lack the agrarian stage of production. That is why they are atypical of the Kaiserstuhl. They can be found more often in communities that do winegrowing on an insignificant scale or not at all and which are situated close to the winegrowing areas.

Winegrowers' Cooperatives

The winegrowers' cooperatives are by far the most important link between the agricultural producer and his suppliers and consumers. Thus, they occupy a central position in the process of integration (see the economic data in Table 21.2). In the Kaiserstuhl, two types may be distinguished:

1. Winegrowers' cooperatives that hold the whole production cycle, starting with the receiving of the grapes up to the commercialization, and which also must accomplish infrastructural tasks;

Table 21.2 Development of the cooperative sector
of the Badish wine industry.[a]

				Year of Vintage				
	1970	1971	1972	1973	1974	1975	1976	1977
Delivery and deposition, million liters	94.5	54.4	84.8	104.2	53.4	89.2	121.9	146.3
Production of Kaiserstuhl and Breisgau regions, (million liters)			22.0	23.6	17.7	20.1	24.8	31.2
= % of the Badish cooperative sector			26.2	22.7	33.1	23.0	20.3	21.0
Central Winery (ZBW) (million liters)			31.8	40.3	18.6	35.4	57.3	67.0
= % of the Badish cooperative sector			37.3	38.7	34.9	40.0	47.0	46.0
business year		1971/72	1972/73	1973/74[b]	1974/75	1975/76	1976/77	1977/78
Sales, million liters Winegrowers' cooperatives		43.3	40.3	36.2	49.4	48.9	52.4	
Central Winery		27.7	28.0	20.9	29.7	29.4	32.1	
Proceeds, million DM Winegrowers' cooperatives			283.0	220.0	313.3	313.2	344.0	
Central Winery			106.6	81.3				
Profit, million DM Winegrowers' cooperatives				173.0[c]	176.0[d]	202.4[e]	206.6[f]	
Central Winery				109.9[c]	100.0[d]	108.7[e]	107.0[f]	118.4[g]
Members of the Badish Winegrowers cooperatives Number		21,361	22,463	23,126[c]	23,631[d]	23,982[e]	24,907	25,500[h]
Investment stock, million DM		81.9	100.0	100.0	120.0	110.4	127.3	135.0[h]
Own capital, million DM		81.3	102.6	102.6	112.2	122.1	133.8	140.0[h]

a according to Jahresbericht des Badischen Genossenschaftsverbandes 1974 ff. - b only ten months because of the change of the business year from 1/9-31/8 to 1/7-30/6. - c in 1973. - d in 1974. - e in 1975. - f in 1976. - g in 1976. - h in 1977. - preliminary data.

2. Winegrowers' cooperatives that fulfill only tasks concerning the organization. The wine making and the commercialization is left to the Central Winery (ZBW), because the quantitative yield and the number of members is below the limit of profitableness of the winery and facilities for wine aging.

It is characteristic of the Kaiserstuhl area that the first type dominates, as opposed to other winegrowing areas in Baden. Only a few winegrowers' cooperatives deliver their total vintage to Breisach, while the others deliver only a certain percentage. Another important task of the winegrowers' cooperatives is commercial regulation. A huge part of the wine is sold by wholesale chains and branches of food shops, which urge a continous supply of wines that show small variations in quality and quantity. Only the storage of wine on a large scale can guarantee such a continuity. Also, the function of decision making is added: the winegrowers' cooperatives not only give agricultural, technical, and economic information to their members, but they directly take part in decisions made on a large, spatial scale (for example, the proceedings of consolidation of vine land). Members of associations concerned with a consolidation procedure and vine-planting cooperatives are not identical with winegrowers' cooperatives, but they never operate without the participation of winegrowers' cooperatives. These tasks clarify the fact that the activities of winegrowers' cooperatives have remarkably changed since their foundation. Previously, they had been self-help organizations of vintages who considered cooperation the only possible solution during the economic crisis after World War I.

The Central Winery of Winegrowers' Cooperatives of Baden (Zentralkellerei Badischer Winzergenossenschaften - ZBW) has made the most distinct integrative efforts toward further concentration of the wine industry. In 1977, it held 67 million liters, or 46 percent of the grapes registered by cooperatives in Baden. As the cooperatives held 88.7 percent of the whole vintage, the ZBW has a share of 41 percent in the total Badish vintage. Thus, the concentration is far more intensive than elsewhere in Germany, where smaller regional vineries and winegrowing estates are the prevailing

types of enterprises. The ZBW is the biggest winery
in Europe, and it has a share of 14 percent in the
marketing of German wines. It was first founded in
1952 by the cooperation of eight winegrowers'
cooperatives. Now, ninety-one winegrowers' cooperatives and one winegrowing estate supply the ZBW.
It has continually expanded its production and
storage capacity, to almost 140 million liters in
1978. During the harvest, up to 3,800 tons of
grapes per day can be registered and pressed. As
the legislation orders all quality wines to have a
detailed distinction according to the place and
site of origin, the sort of vine, the degree of
quality, and the year of vintage, the ZBW offers
more than a thousand different wines.

The grapes are delivered from all winegrowing
areas in Baden, but the individual winegrowing
regions are represented differently: on the Berg-
strasse and in the Markgraflerland there exist
regional wineries, while the Kaiserstuhl region is
characterized by a preponderance of winegrowing and
wine-dealing cooperatives.

Winegrowers' cooperatives and the Central Winery are, in turn, members of the Cooperative
Association of Baden (Badischer Genossenschaftsverband - Raiffeisen - Schulze-Delitzsch e.V.),
which was formed in 1972 by the amalgamation of the
cooperative association (Genossenschaftsverband)
and the farmers' cooperative union (Raiffeisenver-
band). In this present association, cooperatives
having different economic functions are combined
(loan-banks, supply and sellers' cooperatives,
cooperatives for the utilization of milk, of cattle
and meat, of eggs and poultry, of fruit and
vegetables, and, moreover, buyers' cooperatives and
other merchandize and service cooperatives). Their
total number dropped from 1,602 in 1972 to 1,232 at
the end of 1977, because numerous small cooperatives disbanded or amalgamated with neighboring
larger cooperatives. The exceptional position of
the winegrowers' cooperatives and the ZBW within
the association occurs because of the specialization - another crucial characteristic of horizontal
cooperation and of agro-industrial integration.

As far as the infrastructural sphere of scientific information is concerned, a close cooperation
exists, linking all winegrowing enterprises with
the Governmental Institute of Winegrowing in Frei-
burg, which is subordinate to the Ministry of

Nutrition, Agriculture and Environment of Baden-Wurttemberg. This institute promotes the growing of wine with scientific research and practical application.

In conclusion, structural and functional economic areas as well as spheres of conceptual interaction do not coincide with the winegrowing complex. Supplying and marketing areas exist as spheres of activities, varying in intensity from the center to the periphery. Outwards, the relations are open – above all, as far as marketing is concerned. The insignificant export connections give proof of this, although they are not completely lacking. On the other hand, there exist spheres of influence for normative statutes. They are based on jurisdiction and on authoritative competences that have a direct impact on the spatial organization. They add to the spectrum of regionalizational classification.

The Soviet conception of agro-industrial complexes presupposes a spatial identity of structural facts and functional relations, and does not make the distinction between the above-listed regions. Soviet economic geography emphasizes the taxonomic way of thinking in the hierarchy of economic areas – which is easily comprehensible, if one considers the extension and variety of Soviet lands. This fact should be considered as a remarkable contrast to the results of research work derived from the conditions of a free-market economy. Even if the analogous technical development takes place, free-market economy and centrally planned economy will issue different space types and classes of regions. The considerations of the production process suggests that each of the production levels forms its own spatial connection. Their total integration (spatial) in an agro-industrial complex seems to be extremely difficult.

Therefore, it may be more reasonable to think of the agro-industrial complex in its special form (for example, as a wine industry complex) as an economic or agrarian formation. Thus, agro-industrial complexes are not real spatial units that may be delimitated from each other, but are types of regions that are defined economically and which overlap and complete each other.

NOTES

1. J. Stadelbauer, "Der Weinbaukomplex Kaiser-stuhl," Regio Basiliensis (1978): 143-171.

2. (a) M. I. Kozyrya et al. (eds.), Pravovoye polozheniye argrarno-promyshlennych pred-priyatiy i ob-yedineniy (Kishinev 1974).

(b) A. S. Negru-Bode, Agrarno-promyshlennoye kooperirovaniye v SSSR; (Moscow, 1975).

(c) D. V. Valovoy, Mezhkolkhoznaya i gosu-darstvenno-kolkhoznaya kooperatsiya; (Moscow, 1977).

(d) J. Stadelbauer, "Horizontale und vertikale Kooperation in der sowjetischen Agrarwirtschaft,: Geographiscge Zeitschrift 67 (1979) 211-239.

3. I. Karlyuk, "Formirovaniye i planirovaniye agrarnopromyshlennogo kompleksa," Voprosy ekonomiki 4: (1978) 40-49.

22 The Impact of Modern Crop Production Systems in Hungary: The Case of the Nadudvar Method

P. Eke

Nadudvar, the center of the closed or crop-production system (CPS) under study in this chapter, is located in eastern Hungary. Despite its large population, 9,500, it is a typical agricultural settlement. The village is located on the border of two landscape units of the Great Plain, namely, Hortobagy and the Hajdusag. Owing to its being situated on a border, the conditions of soil are varied and, in part, disadvantageous: alkali soil occupies more than half of the fields of the village, while on the rest there are chernozem (black earth) soils and mainly fertile fields.

In analyzing the general conditions of farming, we can consider the relative nearness of the towns to be favorable. Nadudvar is nearly twenty kilometers from Hajduszoboszlo with its almost 25,000 inhabitants and forty kilometers from Debrecen with a population of 200,000. Though Nadudvar has no direct rail connection, the position of its communication can be said to be auspicious. It is connected to the main Budapest-Szolnok-Debrecen-Nyiregyhaza line at two points with the help of a ten kilometer long subsidiary road. The public road connections are good in all directions.

Before 1945, the village was characterized by self-sufficient, enclosed and underdeveloped agriculture. Other features were a relative surplus population, lack of industry, and backward living conditions, in short, economic stagnation.

During the past thirty years, the position of Nadudvar has essentially changed. The agricultural

development has been particularly rapid, and the
economy and population of the village have been
basically transformed. Today, Nadudvar is characterized by a modern, socialist, large-scale agriculture on one of the best collective farms of
Hungary.

The development of Nadudvar is interesting and
instructive in that the village has emerged from
its earlier situation and become outstanding
through the continuous self-development and innovative capacity of its agricultural cooperative.
Industry has not been developed here; hence, the
village has not become an industrial territory or
center. Nor have the general conditions of its
economy been different from those of other villages. Only the agriculture of the village has
changed radically. Nadudvar has mobilized its own
resources and found progress in agricultural development. The economic position of the village is now
firm, and Nadudvar has acquired a name through its
agrarian economy.

It was the Voros Csillag (Red Star) Producer's
Cooperative of Nadudvar that gave the first impulse
to development, and its role has been determinant
ever since. The cooperative has become the largest
in the country in terms of gross production. The
Voros Csillag was the initiator of the Nadudvar
crop-production system, now the greatest CPS in
Hungary. As a system center, Nadudvar has intensive
economic-territorial connections and it is a national center of agricultural organization, management, and advisory service.

Nadudvar always seeks the possibilities of
further advancement, as is proven by the fact that
it is also the center of the first agro-industrial
association in Hungary. The agro-industrial association, which is vertically integrated, best shows
the innovative capacity of Nadudvar, since in 1980
there are only four such organizations in Hungary.

THE VOROS CSILLAG COOPERATIVE

The Voros Csillag Cooperative was formed in 1950 on
an area of merely 105 hectares. In the 1950s there
were several cooperatives in the village, but it
was the Voros Csillag that gradually became the
leader, and the other cooperatives slowly merged
with it. At present it is the only large farm of
the village and is one of the largest estates of

the country, comprising 17,000 hectares. The number
of its employees was exactly 3,000 in 1980. The
present size of the cooperative was established
around 1960, and this territorial size made the
start and advancement of the industrialization of
agriculture possible. The estate has been suitable
from the 1960s to the requirements of present
technology, and this advantageous situation was and
has been utilized by Nadudvar.

An important feature of the cooperative is
that, during its development over almost thirty
years, it has had to face no economic crises and
has been characterized by steady economic growth.
Its emergence and its production level (significantly above the national average) cannot be explained by natural factors. The conditions of its
soil are represented by the structure of the
utilization of the fields: 58 percent of the area
is arable land and 38 percent is grazing land,
alkali soil. The intensive land use forms are
vineyards and orchards. In addition to the unfavorable conditions of soil, lack of precipitation is
characteristic of this part of the country; the
annual average of precipitation is merely 520 to
540 millimeters. On the other hand, thermal conditions and the amount of sunshine can be considered
to be advantageous. On the whole, natural conditions are of an average character, and they are
unfavorable on one-third of the land of the cooperative.

The steady growth of the Voros Csillag was due
to various factors. The most important factors
were:

- Conformity to the natural conditions;
- A working discipline above the average from
  the very beginning;
- Continuous high standards and professional
  firm management;
- A sensitivity to innovations.

The final reasons for the emergence of the
cooperative are its innovative capacity, enterprise, a reasonable estimation of local and national possibilities, and quick reactions. The management had been in close contact with the members
from the beginning. After the first successes, the
production experiences were accumulated and innovative capacity became a major characteristic feature
of the cooperative.

As early as 1953, shortly after its formation,
Nadudvar introduced a new agricultural organizational form. This was an incentive method of
distributing labor and income, and it has spread in
Hungary as the so-called Nadudvar system. It increased the monetary interests of the members. This
method was applied for about fifteen years, and it
had a significant effect upon the steady economic
growth. Even later, the cooperative searched for
new features in all fields of economy, and a
continuous capacity for self-renewal has become a
characteristic feature of Nadudvar.

At present the Voros Csillag is a modern
socialist, large-scale agricultural enterprise. It
can be described as one having a many-sided,
high-standard, and well-balanced estate with a
large production scale. Tis chapter describes the
cooperative's main economic activities and attempts
to delineate the changes that have resulted from
the use of modern crop production systems.

Animal breeding is also a many-sided and
high-standard branch. There are two types of animals raised; sheep and turkeys. In order to
utilize the pasture land of more than 5,500 hectares rationally, sheep are bred here systema-
tically for a very long time. Twenty percent of the
pasture land is irrigated. In 1977 there were
11,000 sheep. Regarding turkey breeding, Nadudvar
is a production system center in which the qualitative development is continuous. It exports turkeys
as far away as Canada.

Nadudvar is one of the best-mechanized large
farms in Hungary. Also, it has its own building
section, which has completed such buildings as a
new slaughterhouse and a meat-processing plant as
well as buildings and reconstructions in the village. Nadudvar has carried out considerable processing of products for a long time, with meat and
milk being the most important products. In 1978 the
largest investment of the cooperative was completed
- the slaughterhouse and meat-processing plant. Now
the slaughtering and processing of as many as
80,000 to 100,000 swine and 2,000 cattle annually
is possible. In marketing, also, the activity of
the cooperative is extensive. At present there are
eleven commercial units (shops and restaurants)
operated and owned by Voros Csillag.

The cooperative of Nadudvar illustrates that
significant building, industrial, and commercial
activities can be transplanted into socialist

large-scale agriculture. In 1977, one-third of the cooperative's gross production value derived from nonagricultural activity, nearly 40 percent of the returns came from marketing, and 45 percent of all employees were engaged in nonagricultural work.

The farming success of the cooperative is illustrated well by the indexes of value: the production value of the farm in 1977 was 1.1 billion forints, the productivity per land was 177 percent higher than the national average, and the productivity of labor surpassed the national average by 85 percent.

The high standard of farming can be illustrated by the structure of occupations: only 55 percent of the employees do agricultural work proper; 125 persons have higher degrees, and the number of skilled workers is over 1,000.

Mention must also be made of the cooperative's intensive production and regional connections: in 1977 it took part in eleven CPSs as a system leader and in fourteen as a member.

THE CROP PRODUCTION SYSTEM IN NADUDVAR

By the early 1970s, industrialization and integration within the framework of the Voros Csillag ensured a well-developed economic basis for further advances. The greatest advance was the creation of the Nadudvar CPS as a second stage in the process of integration.

In Hungary, industry-like crop production systems began to spread in agriculture in the early 1970s. Most cooperatives have been able to adapt to any innovations, but only a few were able to create systems of their own. Nadudvar was one of the latter. In the Nadudvar crop-production system, it became impossible to advance in the conventional way; it was necessary to take over and adapt the modern foreign technologies and industry-like production organization.

The Formation of the Nadudvar System
And its Role in Hungarian Agriculture

The first crop production system in Hungary was formed in the autumn of 1972 by ten cooperatives, and the first year of its employment was 1973. The majority of the founders were from nearby places, and two were from neighboring counties. The members commissioned the Voros Csillag Cooperative to form and maintain the system center. The conditions of industry-like production were established by the Voros Csillag Cooperative. They established the headquarters of the association for the work of the center, and Nadudvar provided adequate space for buildings (such as the central machine repair shop that was to be built later on).

The Nadudvar center of the crop production system performs its organizing-developing-advisory functions well. Ever since its formation, it has had influence on the member farms. It is a special phenomenon in the case of Nadudvar that the system center is a cooperative and not a state farm. Consequently, the center organizes and manages the production of cooperatives almost exclusively: in 1977 there were only seven state farms among its 318 member farms.

The development of the system is shown in Table 22.1. On the basis of the sizes and number of its member farms, it is now the largest crop production system in Hungary. In regard to the increase in number of member farms, the most significant fact to note is the dramatic growth in the number of farms and in the acreage of the system lands.

The precise name of the system is Nadudvari Kukorica es Iparinovenytermelesi Egyuttmukodes (the Nadudvar Corn and Industrial Plant Cultivation Association), or KITE. The name is clearly misleading, since, in 1977 the industry-like production of as many as eight sorts of field plants was organized from Nadudvar: wheat, corn, sugar beet, sunflower, rice, rape, potato, and soybean.

The role of the system is very important in the agriculture of Hungary (see Table 22.2). The Nadudvar technology is applied on 10 percent of the whole sowing area of Hungary; thus, the undisturbed activity of the system is of primary national interest. The national significance of the system is also indicated by the proportion of the member farms: in 1977 the 318 member estates constituted more than 25 percent of the country's large-scale farms.

Although in 1977 there functioned twenty-one crop-production systems in Hungary, four of them used nearly 80 percent of the system lands of the country. These four estates are of national significance because their organizing-advisory activity

Table 22.1. The development of the Nadudvar crop production system.

Year	Number of farms taking part in the system[+]	Annual growth in the number of farms	Total size of system land (hectare)	Annual growth of system land (hectare)	Average size of the system land of an estate (hectare)
1972	10	–	–	–	–
1973	32	22	15,709	–	491
1974	186	164	106,553	90,844	573
1975	272	86	185,160	78,607	681
1976	284	12	333,712	148,552	1,175
1977	318	34	413,529	79,817	1,300
1978	350	32	519,700	106,300	1,485

[+]Those member farms which produce more than one system plant are counted as one.

Table 22.2. The role of the CPS of Nadudvar in Hungary's plant production
in 1977.

Plant	Share in the total size of Hungary's sowing area (%)	Share in the quantity of Hungary's total production (%)
Wheat	13.5	15.0
Corn	11.0	14.1
Sugar Beet	29.7	35.0
Sunflower	23.0	26.0
Rice	54.9	39.7
Potato	3.6	5.9

includes large parts of the country. The centers of
two of the big estates are villages (Nadudvar and
Babolna-Nagyigmand), while those of the other two
are towns (Baja and Szekszard). Nadudvar and Babolna, as villages, are centers of nationwide systems.
Of the four big farms, Nadudvar is the only place
where a cooperative is the system leader.

Nadudvar manages the CPSs in fourteen counties, in 350 large-scale estates, including 270
settlements, on an area of more than 500,000
hectares. With its agricultural innovative activity, Nadudvar has exerted a considerable effect on
the agrarian development of the area.

Here are some data showing the attachment of
the member estates to Nadudvar in 1977:

- 60 cooperatives had four or more system
plants grown by the Nadudvar system;.
- 67 cooperatives, one-fifth of the member
farms, had a system land of 2,000 hectares or
more;
- 100 cooperatives, almost one-third of the
member estates, applied the Nadudvar industry-
like technology on 50 percent or more of
their sowing area.

The central functions of Nadudvar are carried
out first of all, by the expert staff of the system
center. In 1978 there were 320 employees in the
center and subcenters. Ninety-two of them had
higher-education qualifications and fifty-four had
secondary education. The group of well-qualified
experts is a guarantee of further development and
constant renewal. The central function of Nadudvar
is also supported by the central stock of spare
parts and accessory machines and the central machine repair shop.

The Farming of the System:
Dispersion of Yield Averages

The application of the industry-like technology
resulted in a considerable and continuous increase
in the yield averages of the member farms, and it
definitely contributed to their greater success.
The system average is 10 to 20 percent higher than
the national average, and in several cases (especially in regard to corn and wheat) it reaches high

Table 22.3. Basic data in the crop-production system of Nadudvar in 1977.

System plant	The number of estates growing system plants	Sowing area of system plants (hectares)	Average land of individual member estates (hectares)	Average yield of the Nadudvar system (q/hectare)	Average yield of Hungary (q/hectare)
Wheat	194	176,703	911	45.2	40.5
Maize	222	140,424	633	59.6	318.7
Sugar beet	161	36,276	225	375.7	318.7
Beet-seed	25	1,600	64	–	–
Sunflower	116	331,784	274	17.3	15.3
Rice	21	15,331	730	9.2	12.7
Rape	19	4,400	232	–	–
Potato	28	3,605	129	217.9	134.5
Soybean	22	3,406	155	14.4	–
Total	318*	413,529	1,300	–	–

*Member farms cultivating several systems plants are counted as one.

316

international standards.

Table 22.3 supplies information on the farming of the system in 1977. The distribution of the system land according to plants is rather uneven. The four leading plants – wheat, corn, sugar beet, and sunflower – occupy nearly 95 percent of the sowing area. It is also significant to note that no major advances have been made in the systemic production of beet-seeds, rice, rape, potato or soybean, even though these are all staples that are necessary for the consumers and the industrial producers alike.

Yield averages are very high for wheat, sugar beet and potato within the system-lands but somewhat lower in other plants when they are compared to Hungary as a whole. Even within the high system averages, however, considerable and constant dispersions can be found. It is a problem not only in the Nadudvar system, but in other Hungarian production systems as well. To a certain extent, the dispersion of yield averages in the Nadudvar system is too great and economically unfavorable. Of the member farms in 1977, 30 percent of wheat, nearly 25 percent of corn, almost 40% of sugar beet, and 35 percent of sunflower farms had yields below the national average. Table 22.4 shows the considerable dispersion of the yield average of sugar beet.

The significant dispersion of the yield averages is an explainable, but definitely disadvantageous, characteristic feature of the Nadudvar system. During the period of rapid development, cooperatives with lower standards of production, cultivating on areas with unfavorable or moderate natural conditions, also became members of the system. These estates are mainly situated on the periphery of the Nadudvar system. They cannot make use of the advantages of the system-like farming and they have little hope of following the lead of Nadudvar. It is especially necessary to adapt the technology to the circumstances of these farms.

SUMMARY

The dynamism of Nadudvar lies in its success in developing a modern agriculture base to ensure its continued economic growth. A great deal of the dynamism can be explained by the fact that, as a result of the successes of the collective farm,

Table 22.4   Dispersion of Sugarbeet Yields in the
             Nadudvar System in 1977.

Yield average groups q/hectares	Member estates		Sowing area	
	No.	%	hectare	%
0 - 318.7	62	38.5	9,574	26.4
318.8 - 360.0	25	15.5	7,053	19.4
360.1 - 390.0	23	14.3	4,546	12.5
390.1 - 430.0	24	14.9	5,798	16.0
430.1 -	27	16.8	9,305	25.7
Total	161	100.0	36,276	100.0

since 1971 the number of the inhabitants of Nadudvar has not decreased. Thus, modern socialist
agriculture may have attractions to keep the populations within settlements. The age and qualification distribution of the population gradually becomes better, and the transformation of the structure of occupation in favor of the indirectly
related agricultural fields is rapid. The transformation of living conditions and the development of
the infrastructure in Nadudvar have also accelerated. An important factor of its long-term development is that Nadudvar as a village has become a
center of special attraction through its modern
agriculture.

23 Conclusion: The Impact of Modernization on Agriculture

Joseph Held

The concept of modernization is controversial by its very nature. From its inception at about the time of the European Renaissance, every generation has considered itself more "modern" than the one before. In fact, the men of that age regarded an entire historical period, the Middle Ages, as pre-modern and as something to be deplored and forgotten. The notion that each succeeding generation somehow knew more than the one before was embodied in the idea of "progress," parallelled by the idea that we are "older" than the ancients and are, thus, entitled to regard our forebears with benevolent skepticism, at best, or consider their knowledge with barely disguised contempt, at the worst. August Le Compte's "positivist" development of mankind in three stages reflected such convictions, and these eventually became dominant in the thinking of most leading intellectuals of Europe, shaping the general climate of society.

Thus, it should not be surprising that we, the successors to these conceptions, consider "modernization" as something inherently good or even inevitable, a powerful engine of history that marches on triumphantly and inexorably. Hence, the concepts of modernization and modernity have been closely associated with the notion of progress, with the one being considered almost interchangeable with the other and both being considered desirable.

The problems of clarification began to emerge when answers were sought for such pertinent questions as "What is modernity?" "What is progress?"

"What is modernization?" The debates that emerged around these issues and the amounts of paper and ink that have been spent on them have been enormous. (1) But it seems safe to say that all attempts at arriving at a "definitive" answer to the dilemmas before us have failed. This failure orginated in the equation that has often been drawn between the concepts of progress and modernization, on the one hand, and the assertion that the two processes must go on forever. For if "modernization" would always mean progress, then one would not need to worry about totalitarian dictatorships, which, according to some prominent scholars, could not have come about without modern means of social control. (2)

The term "modernity" has also come to be associated in many instances with single-variable theories regarding the process of complex changes in societies as embodied mainly in industrialization and its consequences, the urbanization of society and the development of mass culture. (3) Looking at modernity from this point of view, the next logical step would seem to be to assert that all other aspects of social development, including such diverse activities as the arts, education, and, yes, agriculture, should be subordinate factors of change in the contemporary world. It would also follow, for those desiring to accept such single-variable theories, that the relative neglect of these subordinate aspects of social activities is not only acceptable, but perhaps even desirable, on the road to the great race for modernity.

Theories, of course, tend to come up against some stubborn, hard facts of real life. Whenever such notions of single-variable theories of modernization have gained credence, their acceptance resulted in the general retardation of social and economic developments. In fact, one of the underlying premises of this concluding chapter is that if the modernization of one specific segment of economic or social activities is stressed over all others, this will result in the continuing backwardness of developing societies as a whole.

If history is ever to teach us anything, it is that in practically all industrial societies, with some very few but notable exceptions, of course, modernization began in the distant past with fundamental changes in all sectors of society - but with a stress on the primary sector, agriculture. Such changes included inventions to speed up and make physically less demanding the jobs that had traditionally used up a great deal of the energy resources of the food-producing segments of society, introducing better methods and techniques of working the land, freeing farmhands for nonagricultural occupations, and finally, political and economic measures to drive off the land those who would not otherwise leave although their jobs had been eliminated.

One could argue that the introduction of modern production methods eventually included mechanization, the use of increasing amounts of chemical fertilizers, and a more efficient organization of work. These are certainly aspects of change that had to originate in the industrial sector. Yet, agriculture began to develop and increase its productivity long before industrialization could contribute its share to the process. The invention of a better horse collar, the discovery of winter fodder crops, the increasing use of power sources such as the windmill, or the invention of the heavy plough did not need highly developed industrial technology to effect the changes in agriculture. Only after these changes had taken root could the rural sphere provide sufficient quantities of food and surplus labor for the secondary sector, industry, and still retain a sufficient number of agrarian workers for food production.

Let us stop here for a moment. A recently deceased distinguished Hungarian economist, Viktor Kulcsar, had published a work entitled The Changing Village (4) in which he suggested that an advanced stage of modernity is reached in any society when 75 to 80 percent of the population live in an urban region and work in industry and the service sectors. Such an urban-centered view of modernity itself may be entirely correct, and it is certainly typical of those who advocate a single-variable theory of modernization. One might even argue further, on this basis, that the classical pattern of modernization in which the primary sector had to provide food first before the urban population and industry could expand, is no longer valid in the second half of the twentieth century. The argument goes on to state that there now exist some highly modernized societies capable of and quite willing to help less-developed people with the means of modernization, although such help always results in detrimental effects upon single, underdeveloped sectors, such as agriculture. However, such arguments miss the point, since, in cases when they apply, the process of modernization usually occurs under duress - internal or external coercion - and the happy recipients of "unselfishly" provided largesse eventually have to pay a heavy price for a lopsided, often artificially propped-up development that is ultimately against their interests.

Be that as it may, what is one going to make of the fact that the transformation of agriculture has truly become a process with its own, seemingly unstoppable momentum? This seems to be a fact of life in our age, one that can be shunted aside only for a short period of time in any given society without incurring serious economic and political dislocations, accompanied by social tensions and even disorders. As indicated above, the elements of transformation include the mechanizations of production, the use of chemical fertilizers, and new organizational techniques. But there is more to the modernization of agriculture than that. The process also entails the creation of modern means of communication and transportation, and the development of agribusinesses and agro-industry. It includes the idea of production for profit and the reorientation of the agriculturists' thinking along rational, problem-solving lines, as well as the mobilization of labor reserves, such as women, and the elimination of rural surplus labor, paradoxical as these last two may sound. Furthermore, modernization is a process that has to be rooted in the organic evolution of a society, since, ultimately, it means the drastic transformation of the family structure of former peasants, their work habits and social customs, and the eventual elimination of separateness between the urban and rural spheres. In short, rural modernization in the second half of the twentieth century has come to mean the elimination of an entire way of life that has existed for many centuries in the villages and settlements of the rural sphere.

But one must be reminded of the obvious historical context of this process; namely, that rural transformation, although it is now happening everywhere in the world, occurs in societies whose historical traditions and experiences vastly differ; consequently, their approaches to modernization should or even must be different. Although a superficial observer should think that, under the guidance of one or another of the currently pre-

vailing ideological strictures, the transformation
of the rural spheres could proceed on the basis of
a relatively ˙few models, a closer look at the
process itself will negate such a notion. In fact,
even in the currently existing political blocs
dominating the world today, agricultural modernization varies to such a degree that it is justified
to discuss individual cases in sui generis while,
at the same time, making allowances for the presence of some common characteristics.

It is, therefore, important to suggest that
the transformation of agriculture has several sets
of factors, one of which is obviously material, and
is hence measurable with already available tools,
while others, for want of a better word, could be
called attitudinal in nature and are less amenable
to exact measurements. The material elements differ
in each case according to several variables, including the availability of resources, both human
and technical. The attitudinal factors are largely
determined by the respective historical traditions
of a given society, and partly by the changing
context of everyday challenges provided by normal
human interactions at all levels of existence. It
is, of course, impossible to generalize as to which
set of factors has primacy in the process of
change. Yet it seems that emphasizing one set over
the other has not yet proven useful in the real
world. Nonetheless, it is clear that there is some
sort of correlation between the material and attitudinal factors, with one sometimes reinforcing, at
other times weakening, the other. This correlation
can easily be shown if we reach back into the
second half of the nineteenth century to illustrate
this point.

After the emancipation of the Hungarian serfs
in 1848, there was widespread expectation among
agricultural experts that the new opportunities
provided for the peasants would spur them on to
increased production, induce them to accumulate
capital, and, in general, lead them to a higher
level of existence. (5) It was hoped that the
freeing of the land from feudal encumbrances and
the lifting of serf obligations from the shoulder
of rural folks would create conditions in which
they would realize that more work meant more income
and a more comfortable existence.

What really happened was explained by a Bavar-
ian agricultural expert, Ditz, who had been sent by
his government to study Hungarian agrarian condi-

tions in the late 1860s. (6) He reported that the peasants, instead of producing more after emancipation, refused to do so, since they did not actually need more goods than they possessed. In fact, the consequences of emancipation included a drop in production levels, since the peasants no longer wished to work on the large estates as they had done before the emancipation – even for wages! They lacked initiative for another reason: they had been so accustomed to receiving orders from overseers that they were reluctant to work on their own. What they did, instead, was look for buried treasure by which to establish their family fortune. (7)

Changes in peasant attitudes in Hungary began to emerge only in the early 1880s, under the combined influence of increased commercial opportunities, a developing crisis of declining agricultural prices, and the coming of age of a new generation of peasants whose memories no longer included personal experiences of serfdom. The gradual modernization of rural life really began at that time in Hungary, a process that has gone on with stops and starts ever since. (8)

It is obvious that there are a great many differences between that historical period and our own, and that the Hungarian example may not serve as a model of the influence of peasant attitudes on modernization in other countries. The process of modernization in the twentieth century obviously differed from that of the nineteenth century, mainly because of a higher level of scientific and industrial advancement about which the men of previous centuries could not even dream. This involved everywhere the introduction of new tools, long-range planning, and advanced psychological methods to advance the cause of rural transformation. The role of governments in the process has also grown until, in the 1970s, it has become one of the most decisive elements of the transformation process.

It has also been true everywhere that primary consideration in the twentieth century has been given to the development of industry; agricultural progress remained an appendage of industrial developments. This resulted in the neglect of the human factor to the detriment of the entire modernization process. This is one generalization to which some good exceptions can be found. Not so strangely, perhaps, it is in societies where the exceptions were present that modernization has been a more balanced and a more successful process.

In short, then, for modernization to be successful, the modernizers must try to act in a balanced manner. Neglecting the human factor and the human cost by engaging in a speedy industrialization of the entire system will lead to the ruin of agriculture and the ruin of the countryside. Modernization of agriculture is most likely to be successful when a new technology can be utilized to assist the population in changing, not merely the bases of production, but the attitudes of the producers. If such a modernization attempt is carried out at the expense of the people of the countryside, production is bound to suffer and remain stagnant; the stick cannot substitute for the carrot. On the other hand, the modernization of the countryside that takes place with the concomitant modernization of attitudes, offering real incentives to the people, also creates a modernization process whose effects are beneficial for the entire polity as well.

NOTES

1. Some of the more interesting works include; Cyril E. Black, The Dynamics of Modernization. A Study in Comparative History (New York: Harper & Row 1966); Barrington Moore, Jr., Social Origins of Dictatorship and Democracy (Boston: Beacon 1966); Daniel Lerner, The Passing of Traditional Society (N.Y.: Free Press, 1964); Myron Weiner, Modernization: The Dynamics of Growth (New York: Basic Books, 1966); Alex Inkeles and David H. Smith, Becoming Modern (Cambridge, MA: Harvard University Press, 1974); Robert Redfield, The Little Community; Viewpoints for the Study of a Human Whole (Chicago: University of Chicago Press, 1955); Albert O. Hirschman, The Strategy of Economic Development (New Haven, Yale University Press, 1958); W.W. Rostow, Stages of Economic Growth (Cambridge, MA:, MIT Press, 1961), and many others.

2. See Hannah Arendt, The Origins of Totalitarianism (N.Y.: Harcourt & Brace, 1951), pp. 392-437; Carl J. Friedrich, (ed). Totalitarianism (Cambridge, Harvard University Press, 1954)

3. See Dean C. Tipps, "Modernization Theory and the Comparative Study of Societies; A Critical Perspective," Comparative Studies in Society and

History 15 (1973): 201-9.

4. Viktor Kulcsar, A valtozo falu (The Changing Village) (Budapest: Gondolar, 1976).

5. See Antal Voros, "The Emancipation and Its Impact on the Peasantry in 1848-1850," in Joseph Held, ed., The Modernization of Agriculture: Rural Transformation in Hungary, 1848-1975 Boulder, New York: East European Quarterly Columbia, 1980), pp. 24-30.

6. Heinrich Ditz, Ungarische Landwirtschaft (Pest: N.P., 1869).

7. Voros, "The Emancipation," 72-7.

8. Ibid., 91-8.

Index